T0222745

DIGITAL FOOD CULTURES

This book explores the interrelations between food, technology and knowledge-sharing practices in producing digital food cultures.

Digital Food Cultures adopts an innovative approach to examine representations and practices related to food across a variety of digital media: blogs and vlogs (video blogs), Facebook, Instagram, YouTube, technology developers' promotional media, online discussion forums and self-tracking apps and devices. The book emphasises the diversity of food cultures available on the internet and other digital media, from those celebrating unrestrained indulgence in food to those advocating very specialised diets requiring intense commitment and focus. While most of the digital media and devices discussed in the book are available and used by people across the world, the authors offer valuable insights into how these global technologies are incorporated into everyday lives in very specific geographical contexts.

This book offers a novel contribution to the rapidly emerging area of digital food studies and provides a framework for understanding contemporary practices related to food production and consumption internationally.

Deborah Lupton works across the Centre for Social Research in Health and the Social Policy Research Centre at UNSW Sydney, and leads the Vitalities Lab. Her latest authored books are *The Quantified Self* (2016), *Digital Health* (Routledge, 2017), *Fat*, 2nd edition (Routledge, 2018) and *Data Selves* (2019).

Zeena Feldman is Lecturer in Digital Culture at King's College London, where she leads the Quitting Social Media Project. Her work examines intersections between online communication, technology and everyday life, and has appeared in *Information, Communication & Society, TripleC* and *OpenDemocracy*, and on BBC Radio 3 and 4.

Critical Food Studies

Series editors: Michael K. Goodman, *University of Reading, UK* and
Colin Sage, Independent Scholar, *The American University of Rome, Italy*

The study of food has seldom been more pressing or prescient. From the intensifying globalisation of food, a world-wide food crisis and the continuing inequalities of its production and consumption, to food's exploding media presence, and its growing re-connections to places and people through 'alternative food movements', this series promotes critical explorations of contemporary food cultures and politics. Building on previous but disparate scholarship, its overall aims are to develop innovative and theoretical lenses and empirical material in order to contribute to – but also begin to more fully delineate – the confines and confluences of an agenda of critical food research and writing.

Of particular concern are original theoretical and empirical treatments of the materialisations of food politics, meanings and representations, the shifting political economies and ecologies of food production and consumption and the growing transgressions between alternative and corporatist food networks.

Ecology, Capitalism and the New Agricultural Economy
The Second Great Transformation
Edited by Gilles Allaire and Benoit Daviron

Alternative Food Politics
From the Margins to the Mainstream
Edited by Michelle Phillipov and Katherine Kirkwood

Hunger and Postcolonial Writing
Muzna Rahman

Digital Food Cultures
Edited by Deborah Lupton and Zeena Feldman

For more information about this series, please visit: www.routledge.com/Critical-Food-Studies/book-series/CFS

DIGITAL FOOD CULTURES

Edited by
Deborah Lupton and
Zeena Feldman

Routledge
Taylor & Francis Group

LONDON AND NEW YORK

First published 2020
by Routledge
2 Park Square, Milton Park, Abingdon, Oxon OX14 4RN

and by Routledge
52 Vanderbilt Avenue, New York, NY 10017

Routledge is an imprint of the Taylor & Francis Group, an informa business

© 2020 selection and editorial matter, Deborah Lupton and
Zeena Feldman; individual chapters, the contributors

British Library Cataloguing-in-Publication Data
A catalogue record for this book is available from the British Library

Library of Congress Cataloging-in-Publication Data
A catalog record has been requested for this book

ISBN: 978-1-138-39254-0 (hbk)
ISBN: 978-1-138-39259-5 (pbk)
ISBN: 978-0-429-40213-5 (ebk)

Typeset in Bembo
by codeMantra

CONTENTS

List of contributors   *viii*

Acknowledgements   *xii*

1   Understanding digital food cultures   1
    Deborah Lupton

PART 1

Bodies and affects   **17**

2   Self-tracking and digital food cultures: surveillance and
    self-representation of the moral 'healthy' body   19
    Rachael Kent

3   Carnivalesque food videos: excess, gender and affect
    on YouTube   35
    Deborah Lupton

PART 2

Healthism and spirituality   **51**

4   You are what you Instagram: clean eating and the symbolic
    representation of food   53
    Stephanie Alice Baker and Michael James Walsh

5 Healthism and veganism: discursive constructions
of food and health in an online vegan community 68
Ellen Scott

6 Working at self and wellness: a critical analysis of
vegan vlogs 82
Virginia Braun and Sophie Carruthers

PART 3
Expertise and influencers **97**

7 A seat at the table: amateur restaurant review bloggers and
the gastronomic field 99
Morag Kobez

8 I see your expertise and raise you mine: social media
foodscapes and the rise of the celebrity chef 114
Pia Rowe and Ellen Grady

9 'Crazy for carcass': Sarah Wilson, foodie-waste femininity
and digital whiteness 129
Maud Perrier and Elaine Swan

PART 4
Spatialities and politics **145**

10 Are you local? Digital inclusion in participatory
foodscapes 147
Alana Mann

11 Visioning food and community through the lens
of social media 162
Karen Cross

PART 5
Food futures   **177**

12 Connected eating: servitising the human body through
   digital food technologies   179
   Suzan Boztepe and Martin Berg

13 From Silicon Valley to table: solving food problems by
   making food disappear   193
   Markéta Dolejšová

Index   *209*

CONTRIBUTORS

Editors

Deborah Lupton is SHARP Professor in the Faculty of Arts & Social Sciences, UNSW Sydney, Australia, working in the Centre for Social Research in Health and the Social Policy Research Centre and leading the Vitalities Lab. She is the author/co-author of 17 books, the latest of which are *Digital Sociology* (Routledge, 2015), *The Quantified Self* (Polity, 2016), *Digital Health* (Routledge, 2017), *Fat*, 2nd edition (Routledge, 2018) and *Data Selves* (Polity, 2019). She has also edited/co-edited six books. Deborah is a Fellow of the Academy of the Social Sciences in Australia and holds an Honorary Doctor of Social Science degree awarded by the University of Copenhagen. Her blog is 'This Sociological Life'.

Zeena Feldman is Lecturer in Digital Culture in the Department of Digital Humanities, King's College London. Her research investigates intersections between communication, technology and everyday life. She is especially interested in how digital technologies impact understandings and performances of traditionally analogue concepts – for instance, friendship, work and wellbeing. Her work has appeared in *OpenDemocracy, Information, Communication & Society, TripleC* and *Cultural Policy, Criticism & Management Research*. She is co-editor of *Art & The Politics of Visibility* (Bloomsbury, 2017) and author of the forthcoming monograph *Belonging in a Social Networking Age*.

Contributors

Stephanie Alice Baker is a Lecturer in Sociology at City, University of London, UK. Her work explores technological developments as they relate to the formation of knowledge, identity and relationships, particularly online.

Her current research is centred around questions about how digital technologies are changing the way we manage our health. She has recently completed her second book on *Lifestyle Gurus* (Polity, 2019), which explores how digital communication has altered how authority and influence are achieved online, particularly in the domain of health and wellness.

Martin Berg is Associate Professor of Sociology and Media Technology at Malmö University, Sweden. His research engages with processes of datafication and digitalisation through ethnographies with devices and systems. He has published widely on topics such as online ethnography, wearable technologies, self-tracking and digital health. He is the co-author (with Vaike Fors, Sarah Pink and Tom O'Dell) of *Imagining Personal Data: Experiences of Self-Tracking* (Bloomsbury Academic, 2020).

Suzan Boztepe is Senior Lecturer in the Department of Computer Science and Media Technology at Malmö University, Sweden. Her research interests include design as a driver of social and organisational change, generating economic and user value by design, service design and the strategic impact of design in organisations. She is the co-editor (with Clive Dilnot) of John Heskett's posthumous book, *Design and the Creation of Value* (Bloomsbury, 2017), compiled from his unpublished drafts on economic theory and design.

Virginia Braun is a Professor in the School of Psychology, The University of Auckland, Āotearoa/New Zealand. As a critical feminist psychologist, she is interested in sociocultural representation and information, and possibility and practice for individuals. Her work is on gendered bodies, sex/uality and health, including 'healthy eating' as contemporary practice. She also writes about qualitative research, and is co-author (with Victoria Clarke) of *Successful Qualitative Research* (Sage, 2013), and co-editor (with Victoria Clarke and Debra Gray) of *Collecting Qualitative Data* (Cambridge University Press, 2017). She is currently writing a book on thematic analysis (with Victoria Clarke). She tweets @ginnybraun

Sophie Carruthers is currently working in a research and insights analyst role at a New Zealand media company. She is interested in critical psychology, predominantly around health. She graduated from The University of Auckland with honours in Psychology, where she undertook the research outlined in her co-authored chapter in this volume for her honours dissertation.

Karen Cross is a Senior Lecturer in the Department of Media, Culture and Language at the University of Roehampton, UK. She researches widely on contemporary forms of visual social media, exploring their role in new mediations of memory, citizenship, community, as well as performances of creative labour in the digital age.

Markéta Dolejšová is a design researcher exploring social and ethical implications of digital technology in everyday life contexts, with a specific interest in food and health issues. A majority of her work takes the form of workshop-like experiences that she documents through qualitative ethnographic methods. Her research contributes to the critical discourses in Human–Computer Interaction (HCI), Interaction Design and Food Studies. She received her PhD in Interaction Design at the National University of Singapore (NUS) and is currently affiliated as a Postdoctoral Research Fellow at Charles University in Prague, Czech Republic.

Ellen Grady is a Communications Practitioner and Assistant Director at the Australian Public Service. She has a Master of Strategic Communication which was awarded through the University of Canberra. She is keenly interested in methodologies employed to communicate with influence across different media platforms. Her research interests include the role of mass media in shaping public perceptions, representation of women across media platforms, and 'feminine' leadership.

Rachael Kent is a Teaching Fellow in Digital Media and Culture in the Department of Digital Humanities, King's College London, UK. Her research and publications are concerned with exploring the intersection between health, technology and the body, in particular the role of surveillance and representation within these relationships. Her forthcoming monograph, *The Health Self*, examines how the use of self-tracking technologies and social media performativity influence health management in everyday life. She is a frequent speaker on news media, at public events and inter-disciplinary conferences.

Morag Kobez enjoys food so much she has made it her life's work. She has been a food journalist for more than ten years, and is currently the Queensland restaurant reviewer for *Qantas Magazine* and *Qantas Travel Insider*. She recently completed her PhD in Journalism, investigating how digital disruption transformed the role of the food critic. She also teaches Media and Entertainment units in the Creative Industries and Business Faculties at the Queensland University of Technology, Australia.

Alana Mann is Chair of the Department of Media and Communications, Faculty of Arts and Social Sciences (FASS), University of Sydney, Australia, and a key researcher in the University's Sydney Environment Institute. Her latest book, *Voice and Participation in Global Food Politics* (Routledge, 2019), focuses on the communicative dimensions of citizen engagement, participation and collective action in food systems planning and governance.

Maud Perrier teaches Sociology at the University of Bristol, UK. She has written about feminism, motherhood, class, neoliberalism, food and critical pedagogy in journals such as *Gender and Education, Feminist Formations, Sociology,* and *Australian Feminist Studies* amongst others. She has co-edited with Maria Fannin *Refiguring The Postmaternal: Feminist Responses to the Forgetting of the Maternal* (Routledge, 2018). She is writing a monograph called *Politicizing Childcare: Maternal Workers, Class and Contemporary Feminism.*

Pia Rowe is a Research Fellow at the 50/50 by 2030 Foundation, Institute for Governance and Policy Analysis, University of Canberra, Australia. She is also the Academic Editor of the Foundation's gender equality blog, BroadAgenda. Her research interests include inclusive notions of politics and feminism, in particular issues normally considered social and non-political, as well as alternative forms of activism.

Ellen Scott is a PhD student at the University of South Australia. Located within the sociology of food and digital sociology, Ellen's research investigates the symbolic significance of vegan's everyday food choices and their representations of food online.

Elaine Swan is based at Sussex University in the Future of Work research hub. She researches foodwork in all its manifestations, from the domestic domain to community organising through to food labour in the public domain. A critical race feminist, she has published on critical diversity studies, whiteness and therapeutic cultures, and food pedagogies, with books on each of these topics.

Michael James Walsh is an Assistant Professor in the Faculty of Business, Government and Law, at the University of Canberra, Australia. His research interests include the writings of Erving Goffman, cultural sociology, technology and music. A chief dimension of his research involves exploring communication technologies as they relate to and impact on social interaction.

ACKNOWLEDGEMENTS

This book has its antecedents in a symposium entitled 'Digital Food Cultures' convened by Deborah Lupton at the University of Canberra in October 2017. Most of the contributors to the book presented at this symposium and were subsequently invited to revise and expand their presentations for the chapters published here. We thank the University of Canberra for funding and supporting the symposium. Thanks also go to Professor Mike Goodman, who encouraged us to put the current collection together for his 'Critical Food Studies' book series.

1

UNDERSTANDING DIGITAL FOOD CULTURES

Deborah Lupton

Introduction

For centuries, media portrayals have made a major contribution to the generation and reproduction of cultural meanings and knowledges associated with food: including religious tracts, medical journal articles, cookbooks, film and television series, advertisements, news reports and many other media forms. Since the emergence of personal computing, the internet and the World Wide Web in the 1980s and 1990s, digital media have offered a host of novel opportunities not only to represent food cultures, but also to create and share content. Websites, blogs and online discussion forums have been joined by social media and content sharing sites, mobile devices and apps in producing a richly diverse array of portrayals of food preparation and consumption activities. Encouraged by the mobilities and participatory ethos that has emerged in relation to new digital media, people can now use online technologies and apps to nominate their geolocation and show where they are consuming food, rank and rate restaurants, share their dietary and cooking practices and view those of others, engage in food activism, connect with communities of people who share similar dietary beliefs and locate and order foodstuffs and food-related products.

A search on the internet empire Amazon's platform for 'food' returns over 300,000 results, including products available for online sale such as a multitude of cooking, nutrition and diet books, food products such as sauces, oils, tinned foods, sweets and spices, food storage containers and cooking DVDs. The major app stores Apple's App Store and Google Play feature hundreds of food-related apps for mobile devices, including children's games and apps for cooking and meal-planning, health-related apps providing nutrition and additives information, calorie-counting apps and those designed to support food sustainability and ethical eating practices. Restaurant-rating apps and booking and ordering apps for restaurants and fast food outlets are rapidly increasing in number and popularity. Cooking videos constitute a popular genre on YouTube, while images of food are integral to the content uploaded by users of Instagram and Pinterest. Facebook, the most popular social media site globally, offers an expansive multitude of food-related special interest groups, including those dedicated to niche eating styles, weight-loss efforts and food activism as well as for sharing recipes and cooking hints. Some of the more popular Facebook groups, such as Tasty, Buzzfeed Food and Food Network, have tens of millions of followers.

This book is the first to take up the term 'digital food cultures' to examine representations and practices related to food across a variety of digital media: blogs and vlogs (video blogs), Facebook, Instagram, YouTube, technology developers' promotional media, online discussion forums, and self-tracking apps and devices. As our title suggests, the contributors take a sociocultural approach in their analyses of the entanglements of people, food and digital technologies. The affordances of the various digital technologies examined in the book are important attributes. Affordances are the intentions for use that are designed into digital devices and software, or what they invite users to do (Davis & Chouinard, 2016; Nagy & Neff, 2015). These intended uses may be taken up, or alternatively may be avoided, ignored, resisted, challenged or re-interpreted. From a sociomaterialism perspective, affordances can be understood as engendering and distributing capacities for action as part of people's encounters with digital media and devices. Technological affordances come together with human bodily affordances to generate agential capacities (Lupton, 2019a). Affordances differ between devices and software and different communities of users respond to them in different ways.

This book's chapters emphasise the sheer diversity of food cultures available on the internet and other digital media, from those celebrating unrestrained indulgence in food to those advocating for very specialised diets requiring intense commitment and focus. Restricted diets include the paleo diet (involving limiting foodstuffs believed not to be available to Paleolithic humans, such as grains, sugars, most dairy products, vegetable oils and legumes), veganism (in which adherents avoid consuming animal products), and 'clean' eating (a style that focuses on the consumption of foods that are believed to be 'natural' and 'uncontaminated', such as organic foods and those that have undergone minimal processing). Contributors to this book also demonstrate the entanglements between cultures of human embodiment, identity, ethnicity and social relations with digital technologies. The chapter authors are all located in the global North: UK, Australia, New Zealand and continental Europe. While most of the digital media and devices on which they focus their attention are available and used by people across the world, the authors offer valuable insights into how these global technologies are incorporated into everyday lives in very specific geographical contexts. The book therefore simultaneously takes both a local and a global perspective of digital cultures as they operate in relation to food.

Previous research on digital food cultures

Despite the development of a large and lively body of academic literature on the sociocultural dimensions of digital technologies, the popular and ever-expanding streams of digital content concerning food production, preparation and consumption has not received a high level of attention until recently. With the exception of a book on food and social media by Signe Rousseau (2012), academic interest in digital food cultures has had something of a slow start. However, there are encouraging signs of late that attention to this topic is beginning to burgeon. De Solier (2018), Lewis (2018) and Lupton (2018a) have published overview pieces on digital food cultures. Other researchers adopting a sociocultural perspective have investigated the discursive and visual content of food blogs (Dejmanee, 2016; Hart, 2018; Lofgren, 2013), online restaurant review platforms (Kobez, 2018; Vásquez & Chik, 2015; Zukin, Lindeman, & Hurson, 2017), digital food media in the context of body weight and size (Lavis, 2015, 2017; Lupton, 2017a), apps and platforms for food-tracking (Crawford, Lingel, & Karppi, 2015; Lupton, 2018b; Niva, 2017) and food-related hashtags, GIFs and memes (Lupton, 2019b). Studies have found that the 'food selfie', showing food itself or people cooking, preparing to eat or in the act of eating food, is one of the most common form of selfie uploaded to social media – particularly on Instagram (Mejova, Abbar, & Haddadi, 2016; Middha, 2018).

In other research, sociotechnical imaginaries concerning the novel technology of 3D food printing (Lupton, 2017b) and consumers' responses to these food products (Lupton & Turner, 2018a, 2018b) have been investigated. The portrayal and support of specialised eating practices such as veganism (Forchtner & Tominc, 2017; Hart, 2018; Véron, 2016), the Korean phenomenon of *meokbang*, involving public displays of over-eating (Donnar, 2017) and competitive eating (Abbots & Attala, 2017) in blogs, online forums and social media platforms have also received attention. Some studies have investigated the ways in which people use digital media for culinary practices (Kirkwood, 2018), food-related apps (Didžiokaitė, Saukko, & Greiffenhagen, 2018a, 2018b; Lupton, 2018b) or online weight-loss services (Niva, 2017). Researchers have further focused on the ways in which online discussion forums and social media can be used to cultivate pro-anorexia or pro-bulimia practices (Boero & Pascoe, 2012; Cobb, 2017; Ferreday, 2003; Lavis, 2017) or lead to recovery from restricted eating (Chancellor, Mitra, & De Choudhury, 2016; Ging & Garvey, 2018; LaMarre & Rice, 2017). Several previous analyses have identified the gendered dimensions of digital food cultures. These include studies on how food blogging can be a performance of postfemininity (Dejmanee, 2016), the ways in which norms of masculinity are being incorporated into veganism (Forchtner & Tominc, 2017; Hart, 2018), how a liking for chill-flavoured food is gendered on the YouTube interview show Hot Ones (Contois, 2018) and the gendered intersections of fitness, health and eating cultures enshrined in the #fitspo, #healthylife and #eatclean hashtags and calorie-counting apps (Lupton, 2017a, 2018b, 2019b).

'Food porn' is a term commonly used in food cultures to describe aestheticised visual or verbal representations of food that focus intensely on its desire-inducing qualities. The creation of food porn images by consumers has received the attention of several digital media researchers. They have emphasised the ways in which such portrayals have moved beyond professional photography to snaps or videos taken by people with their mobile devices (Dejmanee, 2016; Donnar, 2017; Lavis, 2017; Mejova et al., 2016; Taylor & Keating, 2018). Digital media users have taken up the term in the use of hashtags such as #foodporn and #foodgasm, particularly in social media or curation platforms that focus on visual media, such as Instagram, Facebook, Tumblr and Pinterest (Lupton, 2019b).

Researchers have demonstrated that social media influencers play an important role in drawing attention to food cultures and trends. Social media influencers are people who have accumulated a large following on social media and profit in some way from their prominent profile. Influencers may be established celebrities who go on to build their brands using social media, but also include unknowns, who establish their public profiles from canny (or in some cases, serendipitous) use of social media. These influencers are also sometimes known as 'micro-celebrities'. They tend to become famous to small niche groups of people and construct their social media content strategically to attract and maintain this following (Abidin, 2016; Marwick, 2015). So-called 'lifestyle influencers' who focus on aspects of everyday life such as diet, food preparation, home production of food, food preservation and sustainable food consumption practices have in some cases attracted a large following on their blogs or platforms such as Instagram or YouTube and generated income from spin-off activities such as publishing books. One example is the young Australian woman Belle Gibson, who contended that she had successfully recovered from terminal brain cancer after following the lifestyle practices she subsequently advocated on her blog, social media accounts and a customised app that then became a book released by a major publishing company. Gibson's claims to educate people in how to cure their cancer using dietary practices were eventually discredited, but not before she had enjoyed a lucrative following of her health advice (Lavorgna & Sugiura, 2019; Rojek, 2017).

The use of digital media and devices for food activism and sustainability initiatives has been addressed in several recent publications. These include discussions of the employment of social media by fishermen and women, chefs and food cart operators in the US state of Rhode Island (Pennell, 2016) and by animal welfare organisations (Buddle, Bray, & Ankeny, 2018). Hashtags are frequently used by activists on social media accounts to draw attention to the practices of Big Food companies (Guidry, Messner, Jin, & Medina-Messner, 2015). The edited collection *Digital Food Activism* (Schneider, Eli, Dolan, & Ulijaszek, 2018) includes chapters on a diverse array of digital media employed to facilitate organised efforts to change the food system and seek social justice in relation to food production and provision. These include discussions of the use of Twitter by diabetes activists (McLennan, Ulijaszek, & Beguerisse-Díaz, 2018), Mexican small coffee producers' use of online record keeping and email to document their production and processing methods (Lyon, 2018), hashtag food activism in Australia (Mann, 2017) and activist organisations' employment of barcode scanner apps, wikis, email newsletters, Twitter and Facebook to support ethical consumption practices (Eli, Schneider, Dolan, & Ulijaszek, 2017).

In recent times, an area of human–computer interaction (HCI) has emerged, focusing on the interactions between people, digital technologies and food preparation and consumption. It is sometimes referred to as 'human–food interaction' (Khot, Lupton, Dolejšová, & Mueller, 2017). A key publication in this area is *Eat, Cook, Grow* (Choi, Foth, & Hearn, 2014), an edited collection presenting contributions by numerous HCI researchers working on the intersections between HCI and human–food interaction. The volume included chapters by researchers in design, computing, urban informatics, media and communication, and social science. They address topics relating to how digital technologies could be used to promote initiatives such as food sustainability, urban agriculture, the reduction of food waste, mindful eating, traditional food cultures, precision farming, better nutrition, local food provision, sharing food experiences across distance, and the alleviation of global food insecurity.

Social media influencers can make an impact in political ways as well. Celebrity chef Jamie Oliver is one such influencer. In 2015, Oliver received the highest views for food-related content on YouTube (Jarboe, 2015). He has his own YouTube channel, which had reached over four million subscribers by late-2019. While he is known for his television series, chain of restaurants and cooking-related products, Oliver has also taken overtly political stances on issues such as childhood obesity, taxing sugary drinks and the quality of food served to children in school canteens and consumed by the general public, including making television series devoted to these topics. He has started a self-proclaimed 'Food Revolution', and actively uses Twitter to use this term as a hashtag to further promote his cause (Barnes, 2017). Oliver has therefore been able to build on his popular appeal as a celebrity chef across his television and social media high profiles to engage in alternative food politics.

This book

Several major topics and discursive themes are evident across the chapters published in our collection, building on and extending previous sociocultural research on digital food practices and representations. A range of theoretical perspectives is taken up: principally those by Foucault, Douglas, Bourdieu, feminist new materialism, Goffman and Durkheim. We have grouped the chapters according to themes, but we also emphasise that these are not mutually exclusive, and that there are many overlaps across the chapters. The themes include: bodies and affects; healthism and spirituality; expertise and influencers; spatialities and politics; and food futures.

Bodies and affects

Several chapters in this volume emphasise the affects and moralities that underpin digital food cultures. They demonstrate that binary oppositions such as those between clean/dirty, female/male, controlled/chaotic underpin many of the meanings generated with and through digital food media. These meanings can be associated with affective forces such as pleasure, delight and a sense of community, but also feelings of embarrassment, disgust, guilt, fear and shame.

Rachael Kent's study reported in her chapter focuses on people who self-monitored food consumption for health- or fitness-related purposes. Her research participants were women and men living in the UK, continental Europe and New Zealand who shared information about their practices on Facebook or Instagram. As Kent notes, these self-trackers viewed themselves as conforming to the responsibilised self not only in disciplining and managing their bodies through self-tracking, but also in supporting communities of self-trackers. Affects such as pleasure at achieving set goals were mixed with feelings of embarrassment, shame or guilt in response to not maintaining their strict regimens, concerns about becoming obsessive about self-monitoring or feeling under pressure to present the idealised self or food consumption practices on their social media accounts. The public nature of their goals and self-tracking metrics, when shared on social media, contributed to these affective responses. Kent raises the importance of the 'cheat day' for people who are closely monitoring their food consumption and energy expenditure. As she notes, the cheat day concept allows people who are otherwise engaged in tightly controlled and regulated practices of bodily discipline to relax their efforts occasionally, while continuing to conform to norms of self-management.

Deborah Lupton focuses in her chapter on a particular genre of YouTube food videos that she entitles 'carnivalesque': deliberately designed to be excessive and evoking such affects as greed, desire and disgust. The cheat day concept also features in one of the YouTubers she examines: a young woman who posts content on fitness training routines interspersed with the occasional video in which she spends the day gleefully indulging in otherwise 'forbidden' food. Lupton's other example is the cooking videos that feature on the 'bro'-style Epic Meal Time YouTube channel. Lupton argues that food porn and carnivalesque portrayals both represent the celebration of the multisensory pleasures of food. Carnivalesque portrayals take this celebration even further, leading it into the realm of the extreme and the grotesque, where the intensely carnal pleasures of food are privileged among all other qualities to the point of 'bad taste' (if not literally, then figuratively). This style of digital food culture represents a counter-response to the scientised, rationalised and mechanical concepts of food consumption that are taken up in digital portrayals of practices such as fitspo and weight-loss diets. Simultaneously, however, these videos reproduce sexualised stereotypes of hyper-femininities and hyper-masculinities and surface forceful affective undercurrents of anxiety and ambivalence concerning excessive or 'wrong' food consumption, revealing the fraught nature of contemporary digital food cultures.

Healthism and spirituality

Dietary styles involving restrictions based on ideas about health, nutrition and, in some cases, religious, environmental or ethical choices and the desire to avoid animal cruelty, are burgeoning in the global North (Allen, Dickinson, & Prichard, 2018; Harrington, Collis, & Dedehayir, 2019; Sachdev & Rengasamy, 2018). Digital media offer extensive resources for people wanting to find and share information about and follow these eating styles and provide support to like-minded consumers. Three chapters in this collection focus on restricted diets – one on clean eating, the other two on veganism – and delve further into the relationships between morality, ethical consumption, health, self-responsibility and embodiment. All three chapters make reference to the discourse of 'healthism', which privileges the achievement of good health over other dimensions of food consumption and focuses on individuals' behaviour rather than the broader sociocultural determinants of health states.

In their chapter, Stephanie Baker and Michael Walsh examine the concept of clean eating as it is represented in Instagram images. They show how normative ideals concerning food consumption and embodiment are entangled in Instagram clean eating portrayals. As they note, even if adherents of clean eating on Instagram do not overtly refer to religious or spiritual beliefs, these concepts tacitly underpin many of the discourses and imagery giving meaning to this practice. Baker and Walsh also demonstrate that clean eating discourses and images as they are expressed on Instagram can work as affirmation rituals that establish identity and status among the community of users who engage with hashtags concerning this lifestyle. They show that most of the clean eating imagery examined in their study depicted human bodies, both female and male, rather than the food itself. Clean eating, as this suggests, was integral to norms of bodily appearance: typically, for both women and men, those relating to youthful, conventionally attractive and sexualised fit embodiment.

Ellen Scott's analysis of user contributions on a vegan online discussion forum points to the entanglements of concepts of health with vegan eating practices that involve 'healthism'. Her study found that discussions of the health benefits of a vegan diet were particularly prominent, with the vast majority of forum contributors commenting favourably on how veganism had improved their health or alleviated the symptoms of medical conditions. Like Baker and Walsh, Scott identifies a quasi-religious tone to many narratives on the forum, in which people presented themselves as having converted to veganism and experienced transformational changes for the better, rendering them morally superior to others. Discourses of renewal and rebirth also included references to achieving youthfulness through conversion to veganism. Here again, as was evident in the clean eating posts on Instagram, youth, attractiveness and health were represented as the key benefits of a vegan diet. A vegan diet was portrayed in similar ways to the clean eating diet as it was represented on Instagram, particularly in terms of the assumed purity of vegan foods compared with those excluded in this style of eating.

The chapter by Virginia Braun and Sophie Carruthers also addresses veganism; this time in a study of vlogs produced by practising vegans that have been shared on YouTube. They focus on two major genres of vegan vlogs: 'What I Eat in a Day/Week' and 'My Vegan Story/Journey'. Braun and Carruthers found that the vloggers were overwhelmingly young, white, able-bodied women with lean bodies. As Scott noted in her analysis of posts on a vegan forum, discussions of vegan consumption practices by these vloggers centred on health and wellbeing rather than ethical or environmental reasons for adopting a vegan diet. The vloggers tended to describe vegan food consumption practices as conforming to an instinctive, natural mode of embodiment. Some vloggers evoked religious and spiritual metaphors when describing their bodies and dietary regimens. It is notable that in both the analysis of clean eating by Baker and Walsh, and that of veganism by Braun and Carruthers, the food consumed as part of these diets was also often portrayed in food porn-style tropes: as decadent, delicious, abundant and physically attractive. This mode of representation counters the common stereotype of foods privileged in clean eating and veganism as plain and tasteless, consumed for virtuous reasons rather than for sensory pleasure. The vloggers were therefore able to present themselves as conforming to natural and healthy modes of food consumption while simultaneously freely indulging themselves.

Expertise and influencers

Content-creating platforms such as YouTube – which from the start has encouraged people to 'broadcast yourself' – have vastly expanded the opportunities for non-experts to engage in culinary demonstrations, while restaurant reviewing has been taken over by ordinary consumers using blogging sites or restaurant-ranking platforms. The affordances of these digital media have contributed to an extraordinary shift of food media from the realm of the expert to that of the amateur.

Morag Kobez shows in her contribution that online restaurant reviewing by enthusiastic ('foodie') amateurs has profoundly transformed the gastronomic cultural field. She writes from her own experiences as a professional restaurant reviewer and draws on interviews she conducted with prominent food bloggers. Kobez found that the amateur reviewers she interviewed all upheld similar ethical principles to those espoused by professional reviewers. However, a key difference was that the amateurs tended to employ a discourse of authenticity and ethical appraisal related to their judgements of the provenance of the food they were assessing and its environmental sustainability credentials (e.g., the 'carbon footprint' or 'food miles' associated with the food). In contrast, Kobez's previous research has found that for professional critics, aesthetic engagement with the culinary experience (including factors such as the atmosphere of the restaurant) rather than the provenance of ingredients in meals was considered an integral basis of their reviews. Her study also identifies the intersections between amateur and professional reviewers: she notes that the amateurs often engaged directly with the work of professionals, assessing professionals' work critically and seeking to complement it. Kobez concludes that professional restaurant reviewers continue to maintain a higher status and authority than amateurs, but that amateurs produce markedly different content that complements rather than replaces the work of professionals.

Several contributors to this volume point out that authenticity is a key attribute of blogs and social media content concerning food cultures as content creators attempt to establish a personal brand or reach micro-celebrity status. Sometimes it is people's cooking skills that elevate them to influencer status; while for others, it is their lifestyle, ethical choices and personal food consumption practices for which they gain a following. As Lupton's chapter and Braun and Carruthers' analysis of YouTube videos show, authenticity can be developed using techniques such as speaking directly to camera and offering intimate insights into the presenters' everyday life, including details of the food they prepare and eat. Some people, such as influencer *extraordinaire* Jamie Oliver, combine elements of all these attributes in building their following.

In their chapter, Pia Rowe and Ellen Grady show that for Australian celebrity chef and lifestyle influencer Pete Evans, his personal experiences with adopting the paleo diet come together with his high public profile as an expert chef in his social media profile. Evans' expertise as a chef also combines his interest in promoting the paleo diet through the publication of cookbooks that provide paleo recipes: including one directed at infants' and young children's diets that was highly controversial. While Evans is well-known in Australia for his expertise as a chef, his attempts to promote the paleo diet through cookbooks represents a move from gastronomy into the domain of health and nutrition. Evans sought to leverage his already celebrity status as a chef in his attempts to prove his credibility as a dietary specialist on his Facebook account. He was heavily criticised by nutritional experts for attempting to use his celebrity status to present a nutritional argument that they perceived to be ill-informed.

Maud Perrier and Elaine Swan take a somewhat different angle in their analysis of another Australian lifestyle influencer: Sarah Wilson, of the 'I Quit Sugar' brand renown. Wilson's high media profile was predicated on her successful business building on advocating for reducing sugar consumption to achieve good health and wellbeing. Having sold her 'I Quit Sugar' enterprise, Wilson then moved to address the issue of food waste reduction. Perrier and Swan highlight how Wilson uses her blog to construct a persona that highlights her credentials as an appreciator of tasty and nourishing food, social consciousness and femininity simultaneously – or what they characterise as a 'foodie-waste femininity'. They show that Wilson presents herself and her endeavours in highly classed and racialised ways that attempt to challenge dominant concepts of the consumption of food scraps, discarded food or leftovers as the domain of thrifty housewives, the poor or non-whites, who have little choice but to rely on food that is 'dirty' and 'unhygienic', and therefore 'disgusting'.

Here again, binary oppositions concerning food are central to the cultural meanings of Wilson's digitised activism. Wilson works strenuously to overturn these stigmatised meanings of food waste by presenting herself as a bourgeoise, normatively feminine food appreciator and activist, who is concerned not about saving money but about eating well while simultaneously achieving ethical consumption ideals. As Perrier and Swan point out, Wilson's use of rhetoric and imagery on her blog assumes that her audiences are similarly white, healthy and privileged, seeking to consume rather than discard food waste as an ethical choice rather than economic necessity. In Wilson's blog, the political antecedents and contemporary dimensions of consuming food waste are obscured in favour of a femininised, white, aestheticised ethos: characterised by Wilson herself as 'simplicious'.

Spatialities and politics

The use of digital media to draw attention to food justice and sustainability issues receives further attention in chapters by Alana Mann and Karen Cross. They focus not on individual influencers or micro-celebrities, however, but on the interactions between localised efforts and the social and political contexts and material infrastructures within which efforts to promote local food cultures in inclusive ways operate. The geolocational elements of these initiatives, both authors demonstrate, are important to consider in the context of broader globalised ecosystems of food production and consumption as well as digital technologies and networks. Both contributions also complement Perrier and Swan's chapter by emphasising the entanglements of notions of race/ethnicity and social class with food cultures as they are enacted in these locales and expressed in digital media.

Mann's chapter addresses the use of digital technologies to promote food justice issues. As Mann argues, the greater connectivity between people and new forms of sociality promoted by digital technologies could potentially lead to marginalised voices and their lived experiences receiving greater prominence in public discourses concerning food justice issues. She argues, however, that these forms of communication and sociality do not always result in better inclusivity of marginalised groups or alleviate the social structural causes of food insecurity and poor nutrition. As in other applications of digital technology use, access to food-related initiatives can be denied to those who lack access to the technologies. They consequently lack a voice in discussions and practices around food justice. Mann calls for a better understanding of the ways in which digital and food inequalities are related. She argues that such understanding requires acknowledgement of the historical elements and spatial locations in which food cultures are expressed and reproduced. People from diverse publics need to be offered opportunities to tell their food narratives as part of their participation in food justice initiatives, including using digital media such as customised platforms to make or strengthen community ties and networks.

In her chapter, Cross focuses on the ways in which digital identities and networks were employed in a south London urban regeneration initiative using community-based food markets. Her insights highlight the broader sociocultural and spatial contexts, both digitised and non-digitised, in which digital food cultures are generated and performed. Cross' research attempts to identify how the concepts, politics and practices of emplaced communities such as those in south London are mediated with and through digital media. She discusses two community-based food initiatives in that area of London: the redevelopment of the Brixton Village Market and the West Norwood Feast community market initiative. She notes that social media and relevant hashtags were used by promoters and supporters of the markets to cultivate and disseminate the community feelings to which these food initiatives aspired. Online communities on globalised platforms such as Facebook and Instagram, therefore, intersected with and amplified the community-building efforts that were physically sited in this very specific locale. However, both places have experienced significant sociocultural changes related to gentrification and have hosted activism related to the Black Lives Matter movement. These histories have created community tensions around initiatives such as the community food markets. As Mann's discussion also noted, efforts to use digital media in inclusive ways need to acknowledge the legacies of historical, spatial, raced and classed diversities among publics rather than blithely assume that all social groups living in the area are willing to support such initiatives.

Food futures

Technology developers use their websites and other online venues for promotion (app stores, industry blogs, YouTube videos) to outline and promote the affordances of their products. These marketing practices, which themselves are examples of digital food cultures, are the focus of this volume's chapter by Suzan Boztepe and Martin Berg. They undertake a critical content analysis of how companies who have developed three novel digital devices and services designed to support nutritious food consumption practices have sought to publicise their wares to potential users. Boztepe and Berg examine such media as product descriptions, user guides, sales pitches, blog posts and video clips to identify the technological imaginaries invested in these food-related technologies. As they show, dominant concepts of human embodiment are invested in these promotional media. These include notions of the human body as a machine-like entity and the importance of scientific calibration and personalisation of food consumption to ensure that it functions efficiently and achieves or maintains good health. Such services are designed to encourage an imaginary of food consumption as a rationalised and metricised health practice rather than a source of sensory pleasure.

In her contribution, Markéta Dolejšová provides an analysis from her ethnographic study of people in the tech-hub of Silicon Valley, California, who promote and engage with the Complete Foods diet that uses powered nutrients mixed with water as a meal replacement. The startup ethos of Silicon Valley entrepreneurialism and experimentation pervades the ways in which the Complete Foods diet (which began with the Soylent initiative) has been developed, promoted and taken-up. Advocates of this diet are immersed in digital cultures through their work, and actively use online forums to share their experiences and ideas for how best to incorporate these foods into their routines so as to maximise their effects. They are hackers and experimenters both as a profession and as a personal life-enhancing project. These meal replacement powders, therefore, can be considered to be firmly imbricated within digital food cultures. As Dolejšová explains, this diet originated in Silicon Valley as a way for technological entrepreneurs to maximise their health and productivity and contribute to environmental sustainability and food waste reduction initiatives by using this product as a convenient substitute for solid meals.

This scientific, personalised and rationalised approach to food consumption is similar to that outlined by Kent and Boztepe and Berg in their chapters. Here again, the Complete Foods diet proposes a model of human embodiment which obviates sensual enjoyment of food for a body-as-machine approach. For the self-trackers described by Kent, the technology business entrepreneurs in Boztepe and Berg's chapter, and the Silicon Valley developers working on meal replacement products, food is fuel, and human bodies are the sites of technological experimentation in the quest for optimisation. All three chapters highlight the ways in which novel food technological imaginaries draw on some of the key ideals of the Quantified Self discourse, in which measuring and monitoring aspects of human bodies is viewed as a way of achieving a better self and better life (Lupton, 2016).

Concluding comments

Taken together, the chapters in this book vividly demonstrate the complexities and ambivalences of food cultures as they are expressed and enacted in digital media and using digital devices. They emphasise that dominant moral meanings around food consumption are both upheld and challenged in digital media cultures. The insights into digital food cultures offered by the contributors ultimately demonstrate that digital technologies are central to many contemporary food practices and meanings. They show that food preferences and habits are simultaneously personal and cultural; ethical and indulgent; virtuous and disreputable; progressive and reactionary; traditional and iconoclastic. The affordances of digital technologies can be used to exert tight control over human bodies or to encourage riotous indulgence. Gender norms and dominant ideals of embodiment, self-responsibility and health can be reproduced, exaggerated, or alternatively challenged and transgressed. Disgust for certain types of food (particularly those coded as impure, contaminating or unhealthy) is both reproduced but also playfully engaged as source of entertainment or challenged through practices of recuperation and transgression. These meanings, affective forces and materialities are generated with and through digital technologies in ways that require continuing further cultural analysis beyond the pages of this book.

References

Abbots, E.-J., & Attala, L. (2017). It's not what you eat but how and that you eat: Social media, counter-discourses and disciplined ingestion among amateur competitive eaters. *Geoforum, 84*, 188–197.

Abidin, C. (2016). 'Aren't these just young, rich women doing vain things online?': Influencer selfies as subversive frivolity. *Social Media + Society, 2*(2). Retrieved from http://dx.doi.org/10.1177/2056305116641342.

Allen, M., Dickinson, K., & Prichard, I. (2018). The dirt on clean eating: A cross sectional analysis of dietary intake, restrained eating and opinions about clean eating among women. *Nutrients, 10*(9), 1266. Retrieved from www.mdpi.com/2072-6643/10/9/1266.

Barnes, C. (2017). Mediating good food and moments of possibility with Jamie Oliver: Problematising celebrity chefs as talking labels. *Geoforum, 84*, 169–178.

Boero, N., & Pascoe, C.J. (2012). Pro-anorexia communities and online interaction: Bringing the pro-ana body online. *Body & Society, 18*(2), 27–57.

Buddle, E.A., Bray, H.J., & Ankeny, R.A. (2018). Why would we believe them? Meat consumers' reactions to online farm animal welfare activism in Australia. *Communication Research and Practice, 4*(3), 246–260.

Chancellor, S., Mitra, T., & De Choudhury, M. (2016). *Recovery amid pro-anorexia: Analysis of recovery in social media.* Proceedings of the 2016 CHI Conference on Human Factors in Computing Systems. ACM, pp. 2111–2123.

Choi, J.H.-J., Foth, M., & Hearn, G. (Eds.). (2014). *Eat, Cook, Grow.* Cambridge, MA: MIT Press.

Cobb, G. (2017). 'This is not pro-ana': Denial and disguise in pro-anorexia online spaces. *Fat Studies, 6*(2), 189–205.

Contois, E.J.H. (2018). The spicy spectacular: Food, gender, and celebrity on Hot Ones. *Feminist Media Studies, 18*(4), 769–773.

Crawford, K., Lingel, J., & Karppi, T. (2015). Our metrics, ourselves: A hundred years of self-tracking from the weight scale to the wrist wearable device. *European Journal of Cultural Studies, 18*(4–5), 479–496.

Davis, J.L., & Chouinard, J.B. (2016). Theorizing affordances: From request to refuse. *Bulletin of Science, Technology & Society, 36*(4), 241–248.

de Solier, I. (2018). Tasting the digital: New food media. In K. LeBesco & P. Naccarato (Eds.), *The Bloomsbury Handbook of Food and Popular Culture* (pp. 54–65). London: Bloomsbury.

Dejmanee, T. (2016). 'Food porn' as postfeminist play: Digital femininity and the female body on food blogs. *Television & New Media, 17*(5), 429–448.

Didžiokaitė, G., Saukko, P., & Greiffenhagen, C. (2018a). Doing calories: The practices of dieting using calorie counting app MyFitnessPal. In B. Ajana (Ed.), *Metric Culture: Ontologies of Self-Tracking Practices* (pp. 137–155). London: Emerald.

Didžiokaitė, G., Saukko, P., & Greiffenhagen, C. (2018b). The mundane experience of everyday calorie trackers: Beyond the metaphor of Quantified Self. *New Media & Society, 20*(4), 1470–1487.

Donnar, G. (2017). 'Food porn' or intimate sociality: Committed celebrity and cultural performances of overeating in meokbang. *Celebrity Studies, 8*(1), 122–127.

Eli, K., Schneider, T., Dolan, C., & Ulijaszek, S. (2017). Digital food activism: Values, expertise and modes of action. In T. Schneider, K. Eli, C. Dolan, & S. Ulijaszek (Eds.), *Digital Food Activism* (pp. 203–219). London: Routledge.

Ferreday, D. (2003). Unspeakable bodies: Erasure, embodiment and the pro-ana community. *International Journal of Cultural Studies*, *6*(3), 277–295.

Forchtner, B., & Tominc, A. (2017). Kalashnikov and cooking-spoon: Neo-Nazism, veganism and a lifestyle cooking show on YouTube. *Food, Culture & Society*, *20*(3), 415–441.

Ging, D., & Garvey, S. (2018). 'Written in these scars are the stories I can't explain': A content analysis of pro-ana and thinspiration image sharing on Instagram. *New Media & Society*, *20*(3), 1181–1200.

Guidry, J.D., Messner, M., Jin, Y., & Medina-Messner, V. (2015). From #mcdonaldsfail to #dominossucks: An analysis of Instagram images about the 10 largest fast food companies. *Corporate Communications*, *20*(3), 344–359.

Harrington, S., Collis, C., & Dedehayir, O. (2019). It's not (just) about the f-ckin' animals: How veganism is changing, and why that matters. In M. Phillipov & K. Kirkwood (Eds.), *Alternative Food Politics: From the Margins to the Mainstream* (pp. 135–150). Abingdon: Routledge.

Hart, D. (2018). Faux-meat and masculinity: The gendering of food on three vegan blogs. *Canadian Food Studies/La Revue canadienne des études sur l'alimentation*, *5*(1), 133–155.

Jarboe, G. (2015). *US viewers watch more UK food videos on YouTube than the British!* Retrieved from http://tubularinsights.com/food-videos-youtube/.

Khot, R.A., Lupton, D., Dolejšová, M., & Mueller, F.F. (2017). *Future of food in the digital realm*. Proceedings of the 2017 CHI Conference Extended Abstracts on Human Factors in Computing Systems. ACM, pp. 1342–1345.

Kirkwood, K. (2018). Integrating digital media into everyday culinary practices. *Communication Research and Practice*, *4*(3), 277–290.

Kobez, M. (2018). 'Restaurant reviews aren't what they used to be': Digital disruption and the transformation of the role of the food critic. *Communication Research and Practice*, *4*(3), 261–276.

LaMarre, A., & Rice, C. (2017). Hashtag recovery: #eating disorder recovery on Instagram. *Social Sciences*, *6*(3). Retrieved from www.mdpi.com/2076–0760/6/3/68/htm.

Lavis, A. (2015). Consuming (through) the Other? Rethinking fat and eating in BBW videos online. *M/C Journal*, *18*(3). Retrieved from http://journal.media-culture.org.au/index.php/mcjournal/article/view/973.

Lavis, A. (2017). Food porn, pro-anorexia and the viscerality of virtual affect: Exploring eating in cyberspace. *Geoforum*, *84*, 198–205.

Lavorgna, A., & Sugiura, L. (2019). Caught in a lie: The rise and fall of a respectable deviant. *Deviant Behavior*, *40*(9), 1043–1056.

Lewis, T. (2018). Digital food: From paddock to platform. *Communication Research and Practice*, *4*(3), 212–228.

Lofgren, J. (2013). Food blogging and food-related media convergence. *M/C Journal*, *16*(3). Retrieved from www.journal.media-culture.org.au/index.php/mcjournal/article/view/638.

Lupton, D. (2016). *The Quantified Self: A Sociology of Self-Tracking*. Cambridge: Polity Press.

Lupton, D. (2017a). Digital media and body weight, shape, and size: An introduction and review. *Fat Studies*, *6*(2), 119–134.

Lupton, D. (2017b). 'Download to delicious': Promissory themes and sociotechnical imaginaries in coverage of 3D printed food in online news sources. *Futures*, *93*, 44–53.

Lupton, D. (2018a). Cooking, eating, uploading: Digital food cultures. In K. LeBesco & P. Naccarato (Eds.), *The Handbook of Food and Popular Culture* (pp. 66–79). London: Bloomsbury.

Lupton, D. (2018b). 'I just want it to be done, done, done!' Food tracking apps, affects, and agential capacities. *Multimodal Technologies and Interaction, 2*(2). Retrieved from www.mdpi.com/2414–4088/2/2/29/htm.

Lupton, D. (2019a). Toward a more-than-human analysis of digital health: Inspirations from feminist new materialism. *Qualitative Health Research, 29*(14), 1998–2009.

Lupton, D. (2019b). Vitalities and visceralities: Alternative body/food politics in new digital media. In M. Phillipov & K. Kirkwood (Eds.), *Alternative Food Politics: From the Margins to the Mainstream* (pp. 151–168). London: Routledge.

Lupton, D., & Turner, B. (2018a). 'Both fascinating and disturbing': Consumer responses to 3D food printing and Implications for food activism. In T. Schneider, K. Eli, C. Dolan, & S. Ulijaszek (Eds.), *Digital Food Activism* (pp. 151–167). London: Routledge.

Lupton, D., & Turner, B. (2018b). 'I can't get past the fact that it is printed': Consumer attitudes to 3D printed food. *Food, Culture & Society, 21*(3), 402–418.

Lyon, S. (2018). Digital connections: Coffee, agency, and unequal platforms. In T. Schneider, K. Eli, C. Dolan, & S. Ulijaszek (Eds.), *Digital Food Activism* (pp. 70–88). London: Routledge.

Mann, A. (2017). Hashtag activism and the right to food in Australia. In T. Schneider, K. Eli, C. Dolan, & S. Ulijaszek (Eds.), *Digital Food Activism* (pp. 168–184): Routledge.

Marwick, A. E. (2015). Instafame: Luxury selfies in the attention economy. *Public Culture, 27*(1[75]), 137–160.

McLennan, A.K., Ulijaszek, S., & Beguerisse-Díaz, M. (2018). Diabetes on Twitter. In T. Schneider, K. Eli, C. Dolan, & S. Ulijaszek (Eds.), *Digital Food Activism* (pp. 43–69). London: Routledge.

Mejova, Y., Abbar, S., & Haddadi, H. (2016). *Fetishizing food in digital age: #foodporn around the world*. Tenth International AAAI Conference on Web and Social Media (ICWSM 2016), Cologne. Association for the Advancement of Artificial Intelligence, pp. 250–258.

Middha, B. (2018). Everyday digital engagements: Using food selfies on Facebook to explore eating practices. *Communication Research and Practice, 4*(3), 291–306.

Nagy, P., & Neff, G. (2015). Imagined affordance: Reconstructing a keyword for communication theory. *Social Media + Society, 1*(2). Retrieved from http://sms.sagepub.com/content/1/2/2056305115603385.abstractN2

Niva, M. (2017). Online weight-loss services and a calculative practice of slimming. *Health, 21*(4), 409–424.

Pennell, M. (2016). More than food porn: Twitter, transparency, and food systems. *Gastronomica: The Journal of Critical Food Studies, 16*(4), 33–43.

Rojek, C. (2017). The case of Belle Gibson, social media, and what it means for understanding leisure under digital praxis. *Annals of Leisure Research, 20*(5), 524–528.

Rousseau, S. (2012). *Food and Social Media: You Are What You Tweet*. Lanham, MD: Rowman Altamira.

Sachdev, N., & Rengasamy, G. (2018). Paleo diet – a review. *International Journal of Research in Pharmaceutical Sciences, 9*(2). Retrieved from https://pharmascope.org/index.php/ijrps/article/view/240

Schneider, T., Eli, K., Dolan, C., & Ulijaszek, S. (Eds.). (2018). *Digital Food Activism*. London: Routledge.

Taylor, N., & Keating, M. (2018). Contemporary food imagery: Food porn and other visual trends. *Communication Research and Practice, 4*(3), 307–323.

Vásquez, C., & Chik, A. (2015). 'I am not a foodie…': Culinary capital in online reviews of Michelin restaurants. *Food and Foodways, 23*(4), 231–250.

Véron, O. (2016). From Seitan Bourguignon to Tofu Blanquette: Popularizing veganism in France with food blogs. In J. Castricano & R.R. Simonsen (Eds.), *Critical Perspectives on Veganism* (pp. 287–305). Cham: Springer.

Zukin, S., Lindeman, S., & Hurson, L. (2017). The omnivore's neighborhood? Online restaurant reviews, race, and gentrification. *Journal of Consumer Culture, 17*(3), 459–479.

PART 1
Bodies and affects

2

SELF-TRACKING AND DIGITAL FOOD CULTURES

Surveillance and self-representation of the moral 'healthy' body

Rachael Kent

Introduction

Self-tracking technologies and social media platforms enable multiple ways to monitor and represent the human body. For users, the everyday sharing of self-tracked health-related content on social media can motivate a commitment to personal goals through being accountable to one's community, and through one's performance as a health-optimising role model for participatory audiences. As Kristensen and Prigge (2018, p. 44) highlight: the subject doing the measuring, is also '"delivering" that material to be measured, interpreting the data and acting on these', thus contributing to a continually evolving, involved and expanding relationship with technology, the body, and the parameters and determinants of optimal health. Self-tracking users understand the self through body consciousness. When sharing data on social media, the platform can function as an 'an archive in the process of becoming' (Tifentale & Manovich, 2015, p. 117). Gathering data in relation to goals can offer digital forms and understandings of self through community, via the processes of self-representation and self-improvement (Rettberg, 2018).

This chapter explores how users construct a moral health identity by using self-tracking technologies and sharing on social media, which is simultaneously embodied as a regulatory tool to manage health in their everyday lives. These practices are traced through four stages of performativity and embodiment. First, users perform this moral 'healthy' identity for self and community surveillance purposes. Second, inactivity and physical rest, as well as indulgence in 'treat' and 'unhealthy' foods are made legitimate through the discourse of 'cheat days', and through the reductive conceptualisations of the body as a machine; balancing inputs (unhealthy consumption and diet) with outputs (exercise). Third, through representation of these processes on Facebook and Instagram, users' bodies are moulded, and consumption practices tailored to the desired aesthetics of what is deemed visually pleasing on these platforms. Lastly, the burdens of pervasively tracking and representing the moral 'healthy' body are attended to, identifying the pressures which arise for these individuals as role models, both online and offline for the gaze of their communities. These practices highlight the many different facets involved in performing the self within digital food cultures on social media platforms.

Background

The morals and values ascribed to health and illness are fluid and socioculturally constructed, rather than existing as physical concepts or states of being (King & Watson, 2005, p. 37). As Cederström and Spicer (2015, p. 5) argue, wellness has become not a choice but a 'moral obligation' for health management under digital capitalism and neoliberalism. Moore and Robinson (2016, p. 2776) observe that neoliberalism refers to 'an affective regime exposing a risk of assumed subordination of bodies to technologies'. As a mode of governance, neoliberalism promotes the self-transformation of health, which is achieved through surveillance of the body, health and fitness via 'self-tracking practices [that] are directed at regularly monitoring and recording, and often measuring elements of an individual's behaviours or bodily functions' (Lupton, 2016, p. 2). Regulation through self-tracking parallels 'broader shifts in identity construction, [whereby] people are no longer bound to the inherited guidelines of the past … morality becomes a project to be designed and depicted in relation to others' (Hookway & Graham, 2017, np). Thus, in a neoliberal society, maintaining 'good' health has been ascribed to the individual, demonstrating the shift from a public welfare state responsibility for health towards individualised practices of health self-care (Crouch, 2011; Davies, 2015).

As Cederström and Spicer (2015) argue, guilt plays an important role in commanding the self to be healthy. Individuals who do not adopt these technologies, therefore, may be 'constructed as failing to achieve this ideal and as consequently at fault for becoming ill or contracting a disease' (Lupton, 2012, p. 240). While seventeenth-century religious discourses viewed the development of ill health as an individual lack of self-discipline and intrinsic moral failing, associated with sinful behaviour (Brandt & Rozin, 1997), health is no longer conceptualised as binary to illness. Rather, it has become representative of 'lifestyle corrections' (Leichter, 1997, p. 359), merging behavioural traits and understandings of health with lifestyle-related decisions. Following Foucault's (1986) definition of morality as a process of ethical self-stylisation through care of the self, the 'moral self' can be understood through validation of being 'good enough' in relation to 'health'-related practices.

Previous research has addressed the use of self-tracking technologies in multiple settings: to quantify the productivity of employees' bodies (Moore & Robinson, 2016), provide ethical justification in corporate wellness schemes (Till,

2018), enable diarising and companionship (Rettberg, 2018), and provide a tool to prevent obesity in international policy (World Health Organisation, 2011). The trend towards the widespread adoption of self-tracking practices demonstrates practices of solidarity through philanthropic data sharing (Ajana, 2017), whilst enabling enhanced knowledge of the self and body (Neff & Nafus, 2016). However, this chapter addresses a research lacuna, by examining the use of technologies in representations of health on social media, and how these performed health identities influence health-related behaviours in users' daily offline lives.

Research on social media has frequently centred around the 'selfie' phenomenon (Lim, 2016), as both a privately surveillant and communal performative activity (Tifentale & Manovich, 2015). Elias and Gill (2017) have examined the use of 'beautifying applications' and filters in constructing an ideal body image. This enables the 'hyperbolic construction of "success stories"' (Heyes, 2006, p. 145), denoting a discourse of reincarnation of one's self and body, which is both potentially achievable and always just out of reach. These findings demonstrate the rise of aesthetic self-tracking (Elias & Gill, 2017). Lavis (2017) has explored the representation of pro-anorexia and 'food porn' images as a tool for intimate sociality and the affective production of the everyday. Meanwhile, Talbot, Gavin, Steen and Morey (2017) have identified how the repetition of images of the body ideal provides an often-unrealistic construction of both feminine beauty and masculine strength, leading to a decrease in body satisfaction.

In social media communities, good health has been equated to being 'fit' and more importantly, not being 'fat' (Goodyear, Kerner, & Quennerstedt, 2019), resonating with Lupton's (2017, p. 126) identification that content circulated on social media still plays a 'serious ideological role in stigmatizing and rendering abject fatness and fat people'. This illustrates the still prevalent conceptual relationship between shame and perceived poor health. Also conceived as a lack of health management, this discourse views such behaviour as a threat not only to individual health but also to society. These comparative and competitive practices are conceptualised as 'social fitness', which 'refers to practices of sharing personal data to facilitate motivation and achieving personal goals' (Lupton, 2018b, p. 562). This chapter examines 'health' management in relation to neoliberal moral self-disciplinary discourses, which position the human being as a subject to be worked upon through the performance of specific representations of a moral healthy body.

Methodology

This research aimed to explore the influences of self-tracking technologies and social media upon self-representations of health, as well as everyday offline health behaviours. It involved empirical ethnographic research over a nine-month period with 14 participants who self-selected through a call for participants on Facebook and Instagram: seven women and seven men (pseudonyms used), between 26 and 49 years of age who regularly (daily/weekly) share health- and fitness-related content on these platforms. Located in the UK, continental Europe and New Zealand, the participants ranged from the everyday layperson to those who were dieting or training for marathons or dealing with illness or disease. Ethical approval was granted by King's College London Research Ethics Committee (Ref: LRS-15/16–2156). Each participant completed a guided reflexive diary over three months (two entries per month), returned to the researcher via email on a word document each month. Two interviews were conducted on Skype: pre and post the reflexive diary period. All were recorded and transcribed.

This chapter presents the findings from the critical discourse analysis of the semi-structured interviews and reflexive diary entries, which reflected on their sharing practices. Due to the wealth of data collected, analysis of the visual content is not included. Discourses are related to disciplinary power and play a dominant role in articulating boundaries of ideological regulation, of the body and mind, and of individual desires and behaviours (Foucault, 1986). Critical discourse analysis (CDA) was chosen to provide a detailed critical analysis of participant language use and interpretations of health management via technology use. CDA analyses the causal relationship between discourse and related practices, and broader social structures, to identify the role that power relations play in constructing ideology in language use (Fairclough, 2010). CDA's flexibility further enables its application within different ethnographic methods: written language in the reflexive diaries and verbal language within interviews, which could then be analysed and situated within a sociocultural and political context.

The first participant interview was used to tease out initial reflections around health-related behaviours before undertaking the reflexive diary. Key themes focused on identifying participants' conceptualisations of health, social media, self-tracking, morality, community, and surveillance. Reflexive diaries enable 'autobiographical reflections about the participants' life worlds' (Kenten, 2010, np), rather than the more visible or easily identifiable aspects of health behaviours, by providing a crucial link between the private mind and public affairs (Plummer, 2001). This also enables a reflection on how practices and behaviours, online and offline, change over time. My analysis of the reflexive diaries framed the individual questions for the final interviews.

These methods provided a unique insight into and critical long-term temporal reflection on self-tracking practices from the perspective of users, which is lacking in current digital culture research. The analysis is presented in the following four sections: 'Performing the moral "healthy" subject', '"Cheat days": input versus output discourse', 'Moulding the body and diet to social media aesthetics' and 'Burdens of tracking and representing the body'.

Performing the moral 'healthy' subject

Self-tracking technologies and social media platforms enable certain ways of acquiring knowledge about oneself, as well as performing this knowledge for the surveillance of the community. Initially, participants felt empowered by tracking their consumption practices: in particular, calorie counting from dieting applications and sharing these data on social media:

> It encourages me to be healthier the more I post.
>
> *(Lara, First Interview, 28, F)*

> I've been quite good with my food app, it definitely does make me feel better.
>
> *(Annie, Diary Entry, 28, F)*

The choice architecture and regulatory design tools of these applications and devices 'nudge' and prompt users to make certain health choices (such as lower calorie consumption) and to undertake certain health behaviours (such as increasing exercise routines or running further):

> I'm now feeling the pressure from the [MyFitnessPal] app to do better.
>
> *(Lara, Final Interview, 28, F)*

Visualisations provided by activity recorded on the device/platform provided gratification for their users, while incomplete recordings or empty graphs generated frustration and guilt: 'at stake are the very lenses we use to see ourselves and others' (Neff & Nafus, 2016, p. 11). The participants frequently spoke about pressure to look 'good enough' or a 'certain way', which was usually in comparison to how their peers would perceive their body and health. The fear of negative community comparison and competition ensured participants adopted more extreme exercise behaviours at times (Goodyear et al., 2019). Frustrations were also experienced when misrepresentations of health identity were perceived in offline settings. As Lou explained when she felt concerned about how her peers judged her eating habits:

> I've been trying to relax my attitude towards food, but it's hard when I eat around people as I feel they are judging. But I feel like I am probably to blame for posting all of the healthy pics in the first place, as then I feel I need to live up to it. I bumped into a friend at the train station the other day who was ordering a bacon roll. And she said I wouldn't approve and jokingly apologised.
>
> *(Lou, Final Interview, 29, F)*

Seeing others posting their 'healthy' behaviours can be motivating for the surveying community, but this can also make users who are inactive, or unable to make 'healthier' decisions, feel guilty. This polarisation of behaviours and knowledge worked both ways and encouraged participants' upward comparisons with users viewing shared content (Festinger, 1954). In turn, participants sometimes judged their own consumption and exercise practices as 'unhealthy', or not 'healthy enough', and curated their posts on social media to show achievements and goals met; they idealised visual representations of consumption habits, which reinforced self-betterment discourses of the 'healthy' moral subject:

> My motivations for sharing were to help others and to keep loved ones posted of my progress through this time. It made me feel very accomplished to share my positive experience, and when I was writing it was like a flood gate opening. I felt alive and inspired, lighting up with enthusiasm to share my experiences.
>
> *(Amy, Diary Entry, 27, F)*

> I feel I have a reputation amongst friends for being fit, healthy and strong and therefore feel responsible for sharing my lifestyle in an attempt to inspire, motivate and lift others up with me. Which in turn helps me with my internal goals of health and fitness ... I feel like a role model and therefore I am responsible for influencing others.
>
> *(Annie, Diary Entry, 28, F)*

Many participants regularly discussed wanting to 'give back' to their online and offline community: to guide, support and inspire because of their own perceived health knowledge and education. Thus, representations of 'healthy' living were perceived as an ideological 'gift' of inspiration, resonating with discourses of 'data philanthropy'. Data philanthropy is defined as 'the increasing push for personal data sharing ... for the sake of collective benefit and the ideal of solidarity' (Ajana, 2017, p. 2). Ajana (2017) examines 'data philanthropy' from the perspective of the corporations who develop self-tracking applications: as a discursive configuration of their use and the potential 'positive' health effects on society, generated through the solidarity of sharing personal data on a large scale. As my research findings show, this discourse proliferates through its endorsement not only by the corporations who promote it, but also by users themselves, who conceive their sharing as inspirational for others' health transformation.

'Cheat days': input versus output discourse

A dominant discourse the participants acknowledged was that eating badly or not exercising can be rectified and 'overcome' by exercising harder, minimising calories, eating 'healthier' nutritious food and sharing this process on social media. This discourse reflects neoliberal judgement over the body, which advocates certain healthy lifestyle practices, in particular self-optimising behaviours, subsumed through discourses of morality, competition with oneself, and comparison to earlier healthy/unhealthy practices and bodies (Cederström & Spicer, 2015). Therefore, these applications 'speak to the very core pathology of ... disease; if I do this, then I have to do that' (Gregory, 2013, p. 8). Participants reported feeling 'healthier' from eating 'well' and feeling 'good' about adhering to self-proclaimed goals around diet and sharing.

'Cheat days' were defined as days off both exercising and eating 'healthily', to indulge in 'unhealthy' favourite foods and to rest the body. The participants recognised that exercising and eating 'well' meant they did not feel guilty about eating and sharing about 'unhealthy' foods or consuming alcohol. The discourse of 'cheat days' advocates that the body works like a machine that harnesses an input versus output engine, which prescribes that if you eat 'unhealthy' food or drink (input) you can rectify the extra calories by burning this off (output):

> Today was a full rest/cheat day for sure! Motivated by weeks of being active, eating reasonably well so I figured I deserved a lazy day!
>
> *(Tim, Diary Entry, 34, M)*

> Had a good run and ate well today, did feel able to go out later on without feeling guilty about eating/ drinking later on.
>
> *(Lou, Diary Entry, 29, F)*

> I ate the pasta because I was going for the run. I wouldn't eat pasta on a non-cheat day.
>
> *(Sophie, Diary Entry, 31, F)*

Certain 'unhealthy' (carb-heavy) foods have to be made 'legitimate' to eat not just for pleasure' but for a 'cheat day', as Sophie discussed when sharing photographs of her meals. Sophie further excused this 'cheat' by saying it would fuel and aid her run. For participants, the discourse of 'cheat days' advocates that poor choices surrounding 'health' can be immediately overcome by a lifestyle overhaul (Cederström and Spicer, 2015). The participants perceived the pre-emption of the 'negative' effects of unhealthy behaviours/consumption as rectifiable via intervention, and 'healthy' practices. Over time, the positive feelings the participants initially associated with self-tracking and the regulation of health dissipated:

> My lack of health and fitness activities, including nutrition, has made a massive impact upon my general mood. I'm feeling agitated and very frustrated with it all.
>
> *(Annie, Diary Entry, 28, F)*

> I was actually using My Fitness Pal previously to track my calories. I've decided to stop as it has become too obsessive. I've spent hours on it lying in bed at night deciding what I am allowed to eat the next day, feeling guilty because I was only allowed 1 apple and actually had 2 and it's messed up my daily intake goal. It's ridiculous.
>
> *(Sophie, Diary Entry, 31, F)*

Perceptions of poor self-discipline, and not maintaining 'healthy' consumption behaviours prompted emotionally embodied guilt. This reflects Gregory's (2013, p. 8) argument that self-tracking technologies may have damaging qualities, created through the 'sense of guilt they engender implying defeat when users go "over" their allotted calories and then recommending exercise to make it up'. They also conceived this rectification in reverse; as much as the body can be immediately transformed into a 'healthy' subject through self-discipline, so too can such behaviours be 'undone' by what they perceived as a lack of discipline. This impacted how they conducted their everyday health-related behaviours – for example, concealing certain food purchases for fear of others' judgment:

> After the gym … I went to the supermarket and picked up a ready meal, then I bumped into someone from the gym, I actually felt so guilty, I was so embarrassed about what was in my basket.
>
> *(Lara, Final Interview, 28, F)*

For Lara, her embarrassment about being 'caught' in the act of shopping for a 'cheat meal' ties into discourses of shame around a lack of commitment to certain health practices. The discourse of shame becomes embodied as a parameter for regulating and performing a 'healthy' body, leading to the online and offline performance of a moral subject, who attempts to 'discipline' an 'unruly' body (Gill, 2007, p. 152).

Moulding the body and diet to social media aesthetics

The participants constructed a health identity with which other users within the social media community could perceive and connect. In particular, food had to be tailored to what was thought to be aesthetically pleasing to digital food audiences. This construction was carefully edited to showcase achievements and goals met, and idealised visual representations of consumption habits, which reinforced self-betterment discourses surrounding the 'healthy' moral subject. Participants underlined how they perceived their health identity and how they wanted to represent this on social media. As Sophie explained in relationship to a photograph of her breakfast on Instagram:

> I feel that the image reinforces me being healthy … The picture summed up my goals in terms of the type of food I want to be eating and image I like to portray on social media. Even down to the jar I used to make the breakfast is relevant as there is definitely something satisfying about eating food that's not only good for you but also looks stylish/modern. It puts more pressure on me to not only eat healthy but make food that's going to look 'good' so I can post it. Am I taking my food too seriously? … I do faff around making my food look good before I take a picture.
>
> *(Sophie, Diary Entry, 31, F)*

Participants were satisfied when their food looked 'stylish'; they ensured that colourful ingredients were carefully arranged and presented on fashionable crockery. Time spent shopping for visually pleasing foods and tableware was a necessary part of the representational and consumption process; from buying the food, to preparing the meals, to capturing and sharing the image on social media, to finally consuming the meal. Curating meals that looked enticing whilst also ensuring their nutritional 'value' contributed to the participants' sense of self as 'healthy' and productive individuals, thus reflecting Cederström and Spicer's (2015, p. 7) argument that 'to eat correctly is an achievement, which demonstrates your superior life-skill'.

In contrast, if meals were not deemed attractive, participants felt irritated that they could not share this. In turn, their personal enjoyment of the food was diminished:

> If I've spent ages making food and it's really tasty and then I go to take a picture and it looks bad, I'm like 'oh I can't post it after all that faffing around'. I couldn't post my breakfast this morning, it really irritated me.
>
> *(Lara, Final Interview, 28, F)*

This consideration, therefore, further dictated consumption practices, which prompted the participants to eat visually and aesthetically pleasing ('healthy') food. These practices of preparing, capturing and sharing 'health' self-representations, in particular food, are labour-intensive and time-consuming processes, which convey an attempt to appear 'in the moment'. This individualistic striving for perfection is best understood as entrepreneurial self-work and, more specifically, self-capitalisation concentrated on the visual register (Conor, 2004), effected through consumer regimes of beauty (Gill, 2007). This pre-meditated curation is only deemed 'worth it' when enough likes or positive feedback are received from the community, which can only be achieved when images are 'attractive enough' to post.

Burdens of tracking and representing the body

For the participants, such 'philanthropic' sharing of data and health-related content was framed as educating, inspiring and helping others. Over time, this became fraught with burdensome pressures to continually document a 'best version' of an optimised 'healthy' and moral life, for the benefit of others:

> When friends come to me and say should I eat this I say 'yeah, eat anything you want in balance'. I feel like because I know the pressure, I put on myself, I would never want to contribute that to anyone else feeling like that.
>
> *(Sophie, Final Interview, 31, F)*

> I felt a bit despondent. It feels very far away from me right now. Sometimes looking at the accounts is inspiring and other times can make you feel inadequate.
>
> *(Lara, Diary Entry, 28, F)*

> I wish I hadn't told anyone I was doing this challenge because I feel like I've put too much pressure on myself to look good enough at the end of it and also what food I am seen to be eating. If no one knew about it then they wouldn't be looking at me and expecting me to look a certain way.
>
> *(Lara, Diary Entry, 28, F)*

This drive for individual education to self-manage health is also achieved through adopting multiple self-tracking devices or using many different platforms. Many participants identified engagement with multiple platforms as a contributory factor to existing obsessive relationships with food and health. Knowing how many calories you consumed, that you slept badly or ran slowly does not provide the self-tracker with anything more than that information. This over-simplification without context dehumanises the user and turns them into to a 'good' or 'bad' number or photographic representation, which is restrictive in its inability to provide context related to external lifestyle factors. Upon reflection, all the participants interrogated this process and felt that self-tracking either quantitatively, through biometric data capture applications, or qualitatively through 'selfies' or food photography, felt regulatory and provided self-proclaimed boundaries.

The participants identified that being 'happier' meant not diligently and obsessively regulating food consumption. This perspective resonates with the assertion by Purpura, Schwanda, Williams, Stubler and Sengers (2011 p. 6) that: 'while personal goals are always culturally influenced, the key distinguishing feature … is that users do not get to choose their own viewpoints but are provided with one by designers' as well as by the cultural norms and etiquettes of these data sharing and digital food cultures. A sense of being 'free' was associated with no longer self-tracking health and sharing, which shifted over time in line with the participants' changing lives. Being a consistent inspirational figure and the continual self-regulation needed to self-track and share one's healthy identity and lifestyle became an exhausting process for all the participants at some point during the research period. Interestingly, even though the participants recognised that all social media and self-tracking users must feel this way, they did digitally disengage, as a means of resisting conformity:

> By not sharing I have that power back with me, that I don't succumb to the norm I guess.
>
> *(Fet, Final Interview, 30, M)*

Therefore, the social media norms within these communities perpetuated the idea that sharing is the common practice and not sharing is the deviant behaviour.

A sense of empowerment is achieved by resisting and quitting these persuasive and coercive technologies. The self, therefore, feels liberated from overt disciplinary and regulatory control through technology and surveillance. For most of the participants, from a longitudinal perspective, these sentiments were echoed as their tracking and sharing practices changed over time. Yet, feeling morally 'good' and managing health was considered to be an important priority of everyday life.

Discussion

These findings illustrate how the morality of individuals can be performed through health identity on social media. Over time, self-worth and self-esteem becomes pervasively tied to self-regulation through acquisition (or the lack) of health- and food-related data sharing. The participants became 'subjects of both the normalising gaze of health as well as their own self-surveillance of who to become' (Goodyear et al., 2019, p. 214). They shaped an idealised portrait of a moral healthy identity by showing off representative and idealised personality traits and behaviours to surveying peers. New forms of visualisation and communication emerge from these identity formations (Ruckenstein & Kristensen, 2018) as each individual is made responsible for a new ethics of self-management, obliged to take responsibility for their own 'health', not just to identify and manage their susceptibilities but also to optimise and perform the moral self through management of diet, exercise, and health.

Many participants conceptualised their bodies as engines or machines. In essence, what you put in, you will get out. Reflecting neoliberal self-betterment discourses which advocate that self-regulation and management will transform and revolutionise individual health (Moore & Robinson, 2016). These practices become a 'system of constant registration and constant inspection' (Goodyear et al., 2019, p. 214). These discourses of shame became internalised and embodied by the participants, and conceivably also by other community members viewing this content within these digital food cultures. For the participants, the guilt attached to 'unhealthy' habits became internalised in anticipation of negative perceptions from others (Kent, 2018), both online and offline. Behaving rationally and returning to regulation according to subjectively defined rules meant renouncing the interdictions required to remove or prevent 'bad' and 'unhealthy' behaviours.

This mechanical approach was embodied and interpreted as re-addressing the balance of 'health' through individual self-disciplinary behaviours and legitimate 'cheat days'. These regulations encouraged external pressures to become internally advocated by the participants, through an individual moralism of health, whereby following such regulations advocated 'good' and 'healthy' behaviours, and resistance to or an active choice to disavow the 'nudges' manifested itself in individualised conceptualisations of immorality and the associated dimensions of guilt, shame and embarrassment. These surveillance and tracking practices intrinsically assign an individual moral obligation to preserve one's own health. This discourse of 'shame' and guilt was identifiable within many of the participants' responses and was embodied in relation to 'bad' and 'unhealthy' decisions. This archaic discourse was strongly reflected in the research findings and encouraged the participants to engage with health moralism through the internalisation of shame, in turn encouraging them to perform self-regulatory and self-optimising health behaviours. This ensured that they confined 'health'-related decision-making processes within the parameters of perceived morally 'right' consumption choices, articulated through data collection, interpretation and representation.

By regularly tracking and sharing food and calorie intake, these practices over time became 'obsessive' and a cause of anxiety for many participants, and in turn an embodiment of self-disciplinary rules over consumption, exercise and, broadly, lifestyle. These practices led to 'ridiculous' determinants of what participants were and were not allowed to do and eat, which they interpreted as impacting on their sense of self-worth, generating feelings of being a 'bad' and immoral person. Shame and guilt were attached to poor self-discipline and not sticking to self-imposed 'rules'. Furthermore, feelings of elation and control became synonymous with an adherence to individual parameters of healthy behaviours. However, at both ends of these 'good' and 'bad' spectrums, the participants acknowledged that extreme or overwhelming feelings in relation to perceived success or moral failure as a result of self-surveillance, regulation and discipline were 'ridiculous'. Yet, by often ignoring the inherently important role of human senses and prioritising technological sensors, the technology still held regulatory and moral ruling over how they perceived themselves as 'good' or 'bad' individuals.

These findings reflect research by Kristensen, Lim and Askegaard (2016), which identified how food choices offered insights into the moral character of citizens, and how this may be perceived by others or performed by the individual. This was also intensified by talking extensively about goals to friends, family or colleagues, as well as by posting regularly about these, which ensured there was a front to maintain: a facade of a 'healthy moral self'. In this way, 'living' up to the expectations of others within online and offline spheres was a concern exemplified by many of the participants.

Tailoring and moulding consumption practices to what is visually engaging on social media, reflects processes of mediation which assert that 'institutionally unbounded assemblage ... [produces a] specific subject who is continually dissatisfied about their appearance [or indeed health] and is thus compelled to embark on new regimes of "self-perfectibility"' (McRobbie, 2009, pp. 62–63). These practices of preparing, capturing and sharing health self-representations, particularly food and meals, are labour-intensive despite participants' attempts to appear 'in the moment', as well as self-transformative. These mediated digital capitalist practices play directly into the hands of businesses who profit from these technologies, through the exploitation of 'not only the desire to produce an appropriate type of body ... [but also] the sense of self-development, mastery, expertise, and skill that dieting [as well as other health modifications] can offer' (Heyes, 2006, p. 137). This 'value' is determined and thus acquired by both the representational health data and the participants' sense of feeling 'healthy'.

Conclusion

A desire to be 'healthy' and to prolong our lives binds us to health and to 'experts' or 'influencers', who patent new lifestyles, diets and risks, leaving every aspect of the life course vulnerable to commercial exploitation in the name of health (Rose, 2007). Health expertise works in alliance with individual ethics. These modes of collaboration are problematic in digital (food) cultures because of their gross over-simplification and overgeneralisation of health and the body. Furthermore, this alliance has damaging implications, with regards to over-exercise, misinformation, or poor mental health. Therefore, through the use of these technologies, individual ethics have become somatic, leading to a pervasive moralism of health; using these devices and platforms further entrenches personal failure within discourses of poor self-discipline and poor health self-management, particularly in the context of neoliberalism, whereby individual behaviours and responsibilities are internally practised, embodied and maintained (Cederström & Spicer, 2015; Davies, 2015; Lupton, 2018a).

The role of data sharing technologies in shaping these participants' understandings of their bodies and health highlights the discourses of self-governance and self-discipline that exist within cultures of self-tracking and data sharing. Participants' regulation to adhere to the parameters set by these technologies deserves further attention, especially when considering the pervasive role these technologies play in users' everyday lives. The moral dimensions within dominant health discourses become inherent within self-surveillance practices and the regulatory design of self-tracking apps, devices and the sharing of related content on social media. This continuous reflexive and self-evaluative cycle also incorporates a simultaneous consideration of upholding or amending others' potential judgements of the appearance of one's own body. Being morally 'good enough' must fall in line with specified regulatory frameworks set by the participants, for example, exercising on certain days or eating 'healthily' six days a week. This moralised datafication of health through the measurement of such regulation, either embodied emotionally or through capture on self-tracking devices, enables subjectively determined 'legitimate' times of renounced moral 'codes' and rules; for example, 'cheat days' which allow eating indulgent 'unhealthy' foods or not exercising. This similarly reflects and reinforces input versus output discourses with regards to health management, whereby poor health practices can be overcome by enacting 'healthy' behaviours, and similarly 'healthy' behaviours can be undone by 'unhealthy' consumption or not exercising.

Making sense of one's own body and health through these self-tracking and data sharing processes shifts definitions of health and ill health, activity and inaction, presentation and concealment, surveillance and privacy. The burdens of self-tracking and sharing, and the self-regulation promoted by these technologies can become emotionally detrimental to users' sense of wellbeing, mental and physical health. The participants perceived not maintaining 'healthy' behaviours as a lack of personal self-discipline. This evolved into a moralisation of health, whereby health and lifestyle choices became tied to ethical parameters of 'good' and 'bad' behaviours. This regulatory discourse was deeply embedded in their sense of moral self. To enact certain 'healthy' behaviours was to be a 'good' person. To 'slip out' of regulatory regimes, even due to external or uncontrollable factors in their lives, meant that participants felt deeply ashamed and their perception of self was of a 'bad', undisciplined individual. In turn, self-worth became pervasively tied to data.

Acknowledgements

This research is funded by the European Research Council as part of the 'Ego-Media' project at King's College London: www.ego-media.org/.

Further reading

Ajana, B. (Eds.). (2018). *Metric Culture: Ontologies of Self-Tracking Practices*. London: Emerald.

Fotopoulou, A., & O'Riordan, K. (2016). Training to self-care: Fitness tracking, biopedagogy and the healthy consumer. *Health Sociology Review, 26*(1), 54–68.

Ruckenstein, M., & Schüll, N.D. (2017). The datafication of health. *Annual Review of Anthropology, 46*, 261–278.

References

Ajana, B. (2017). Digital health and the biopolitics of the Quantified Self. *Digital Health, 3*, 1–18.

Brandt, A.M., & Rozin, P. (1997). *Morality and Health*. London: Routledge.

Cederström, C., & Spicer, A. (2015). *The Wellness Syndrome*. London: John Wiley & Sons.

Conor, L. (2004). *The Spectacular Modern Woman: Feminine Visibility in the 1920s*. Bloomington, IN: Indiana University Press.

Crouch, C. (2011). *The Strange Non-Death of Neoliberalism*. Cambridge: Polity Press.

Davies, W. (2015). *The Limits of Neoliberalism: Authority, Sovereignty and the Logic of Competition*. London: Sage.

Elias, A.S., & Gill, R. (2018). Beauty surveillance: The digital self-monitoring cultures of neoliberalism. *European Journal of Cultural Studies, 21*(1), 59–77.

Fairclough, N. (2010). *Critical Discourse Analysis: The Critical Study of Language*. London: Routledge.

Festinger, L. (1954). A theory of social comparison processes. *Human Relations, 7*(2), 117–140.

Foucault, M. (1986). *The Care of the Self. Vol. 3: The History of Sexuality*. New York: Random House.

Gill, R. (2007). *Gender and the Media*. Cambridge: Polity Press.

Goodyear, V.A., Kerner, C., & Quennerstedt, M. (2019). Young people's uses of wearable healthy lifestyle technologies: Surveillance, self-surveillance and resistance. *Sport, Education and Society*, 212–225.

Gregory, A. (2013). Is our tech obsession making anorexia worse? *The New Republic*, 18 December. Retrieved from www.newrepublic.com/article/115969/smartphones-and-weight-loss-how-apps-can-make-eating-disorders-worse.

Heyes, C.J. (2006). Foucault goes to Weight Watchers. *Hypatia*, *21*, 126–149.

Hookway, N., & Graham, T. (2017). '22 push-ups for a cause': Depicting the moral self via social media campaign Mission22'. *Journal of Media and Culture* 20(4). Retrieved from http://journal.media-culture.org.au/index.php/mcjournal/article/view/1270

Kent, R. (2018). Social media and self-tracking: Representing the 'health self'. In B. Ajana (Ed.), *Self-Tracking* (pp. 61–76). Cham: Palgrave MacMillan.

Kenten, C. (2010). Narrating oneself: Reflections on the use of solicited diaries with diary interviews. *Forum: Qualitative Social Research*, *11*(2), np.

King, M., & Watson, K. (2005). *Representing Health: Discourses of Health and Illness in the Media*. Basingstoke: Palgrave Macmillan.

Kristensen, D.B., & Prigge, C. (2018). Human/technology associations in self-tracking practices. In B. Ajana (Ed.), *Self-Tracking* (pp. 43–60). Cham: Palgrave MacMillan.

Kristensen, D.B., Lim, M., & Askegaard, S. (2016). Healthism in Denmark: State, market, and the search for a 'moral compass'. *Health 20*(5), 485–504.

Lavis, A. (2017). Food porn, pro-anorexia and the viscerality of virtual affect: Exploring eating in cyberspace. *Geoforum*, *84*, 198–205.

Leichter, H. (1997). Lifestyle correctness and the new secular morality. In A.M. Brandt & P. Rozin (Eds.), *Morality and Health* (pp. 359–378). London: Routledge.

Lim, W.M. (2016) Understanding the selfie phenomenon: Current insights and future research directions. *European Journal of Marketing*, *50*(9/10), 1773–1788.

Lupton, D. (2012) M-health and health promotion: The digital cyborg and surveillance society. *Social Theory & Health*, 10(3), 229–244.

Lupton, D. (2016). *The Quantified Self*. Cambridge: Polity Press.

Lupton, D. (2017). Digital media and body weight, shape, and size: An introduction and review. *Fat Studies*, *6*(2), 119–134.

Lupton, D. (2018a). 'I just want it to be done, done, done!' Food tracking apps, affects, and agential capacities. *Multimodal Technologies and Interaction*, *2*(2). Retrieved from www.mdpi.com/2414-4088/2/2/29/htm.

Lupton, D. (2018b) Lively data, social fitness and biovalue: The intersections of health self-tracking and social media. In J. Burgess, A. Marwick & T. Poell (Eds.), *The Sage Handbook of Social Media* (pp. 562–578). London: Sage.

McRobbie, A. (2009). *The Aftermath of Feminism: Gender, Culture and Social Change*. London: Sage.

Moore, P., & Robinson, A. (2016). The quantified self: What counts in the neoliberal workplace. *New Media & Society*, *18*(11) 2774–2792.

Neff, G., & Nafus, D. (2016). *Self-Tracking*. Cambridge, MA: MIT Press.

Plummer, K. (2001). *Documents of Life 2: An Invitation to a Critical Humanism*. London: Sage.

Purpura, S., Schwanda, V., Williams, K., Stubler, W., & Sengers, P. (2011). *Fit4Life: The design of a persuasive technology promoting healthy behavior and ideal weight*. ACM CHI Conference on Human Factors in Computing Systems, Vancouver, BC, Canada, May 7–12.

Rettberg, J.W. (2018). Apps as companions: How quantified self apps become our audience and our companions. In B. Ajana (Ed.), *Self-Tracking* (pp. 27–42). Cham: Palgrave MacMillan.

Rose, N. (2007). *The Politics of Life Itself: Biomedicine, Power and Subjectivity in the Twenty-First Century*. Oxford: Princeton University Press.

Ruckenstein, M., & Kristensen, D.B. (2018). Co-evolving with self-tracking technologies. *New Media & Society, 20*(10), 3624–3640.

Talbot, V.C., Gavin, J., Steen, V.T., & Morey, Y. (2017). A content analysis of thinspiration, fitspiration, and bonespiration imagery on social media. *Journal of Eating Disorders, 5* (40). Retrieved from https://jeatdisord.biomedcentral.com/articles/10.1186/s40337-017-0170-2

Tifentale, A., & Manovich, L. (2015). Selfiecity: Exploring photography and self-fashioning in social media. In D.M. Berry & M. Dieter (Eds.), *Postdigital Aesthetics: Art, Computation and Design* (pp. 109–122). New York: Palgrave Macmillan.

Till, C. (2018). Self-tracking as the mobilisation of the social for capital accumulation. In B. Ajana (Ed.), *Self-Tracking* (pp. 77–92). Cham: Palgrave MacMillan.

World Health Organisation (WHO). (2011). *Mhealth: New horizons for Health through Technologies*. Global observatory series, 3. Geneva: World Health Organisation. Retrieved from www.who.int/goe/publications/goe_mhealth_web.pdf.

3

CARNIVALESQUE FOOD VIDEOS

Excess, gender and affect on YouTube

Deborah Lupton

Introduction

Food can be a powerful affective material, assembling with humans to generate forceful vitalities and intensities (Lupton, 1996, 2019; Probyn, 2003). Digital media offer new ways of generating and distributing these affective forces related to food production, preparation and consumption across large audiences (Lavis, 2017; Lupton, 2017, 2018a, 2019). Food has constituted a major preoccupation of internet content creators and their audiences since the emergence of personal computing and the World Wide Web in the late twentieth century. A rapidly expanding plethora of food blogs and online discussion forums forged the way (De Solier, 2018; Rousseau, 2012), followed in the early years of this century by the development of food-related mobile apps, social media and content-sharing platforms (Lewis, 2018; Lupton, 2017, 2018a, 2019). In food-related domains, the affordances for reaching extremely large audiences of social media sites such as Facebook, Twitter, Instagram and Snapchat and content-sharing sites including Pinterest and You-Tube, combine with the compelling force of visual media that can be uploaded, shared, curated and tagged on these platforms (Lupton, 2018a, 2019).

Cultural scholars and researchers are only just beginning to examine and analyse the vast wealth of these images as they pertain to food cultures and foodways; or what Goodman, Johnston and Cairns (2017, p. 161) describe as 'mediated foodscapes'. In this chapter, I examine the ways in which the phenomenon of excessive food consumption and preparation is portrayed on You-Tube. To do so, I draw on two case studies: 'cheat day' videos posted to the channel of Asian-American fitness and health influencer Stephanie Buttermore, and cooking videos on the Epic Meal Time channel made by a group of white Canadian men based in Montreal. While the Stephanie Buttermore and Epic Meal Time channels focus on food, neither of these video genres ascribes to the 'how-to-cook' videos that constitute a large part of food-related videos on this platform (Delgrado, Johnsmeyer, & Balanovskiy, 2014). Buttermore purchases rather than prepares her cheat day food. The Epic Meal Time team's videos demonstrate how they construct their gargantuan meals, but this mostly involves throwing dishes together (in some cases, literally) by stacking or heaping the components rather than by using refined cooking techniques. Nor can either of these video styles be described as conforming to the 'foodie' media genre, which privileges the quality, taste and aesthetic style of food (Cairns, Johnston, & Baumann, 2010). Instead, food is depicted on both channels for its value as a spectacle in terms of its presentation or consumption, involving the distribution of affective forces such as pleasure, greed, desire and self-indulgence but also bordering on inciting the responses of disgust and repulsion. Both types of videos focus on food products or meals that are culturally coded as 'unhealthy', 'bad' or 'junk'. Audiences are invited to enjoy the vicarious pleasure of watching these YouTubers purchase or prepare these foods and then consume them with gusto.

As such, these food videos could be considered one of the more recent ways of portraying 'carnivalesque consumption' (Cronin, McCarthy, & Collins, 2014) of food. Carnivalesque consumption is a practice that directly counters cultural boundaries about what foods are 'good' to eat, valuing excess and loss of control over appetite. Excessive food consumption has long been a carnivalesque tradition, a way of celebrating feast days and holidays and marking them as extraordinary. Histories of the carnivalesque have demonstrated the importance of periodic loosening of carnal restraint during feast days and festivals, generating affective forces of excitement and indulgence in bodily desires as norms and boundaries are transgressed and authorities mocked (Bakhtin, 1984). Carnivalesque ritual food consumption events are often accompanied by periods of fasting, marking them as even more distinct compared with the disciplining of appetite occasioned by fasting (Bakhtin, 1984). In contemporary Western cultures, 'cheat days' for people who are seeking to lose weight or 'eat healthy' have become one such event. The terminology of the cheat day refers to work-out and weight-loss cultures, in which 'cheat' foods, meals or days relate to the consumption of foods that are otherwise forbidden or severely limited. The terminology of the 'cheat' harkens to the idea that indulgence in these foods are a way of avoiding or side-stepping the discipline required of strict diets; if only for an occasional meal or day.

In this chapter, the approach I take views the images of food and its preparation and consumption as more-than-representational, conveying meaning beyond words. The participatory engagements encouraged by social media and YouTube offer people the opportunity to engage in counter-normative practices and representations. Humans–food–digital technologies–place assemblages are formed and distributed to a potentially vast audience. The technological affordances of YouTube allow audiences to respond by liking, commenting on or sharing the videos and subscribing to the channel to keep up with new content. A diverse range of bodies and embodied practices can be found on YouTube, including those that resist or transgress idealised concepts of the highly regulated, health-obsessed and disciplined body (Lavis, 2017; Lupton, 2017, 2018b, 2019).

To conduct the analysis, I viewed a range of videos on both the Stephanie Buttermore and Epic Meal Time YouTube channels that were available in late 2018. In my analysis, I think with feminist new materialism theory to consider the ways in which the content of the videos invites audiences to respond in certain ways, and the discourses and practices on which the video-makers draw in their attempts to generate these forces, as well as some of the responses by viewers left in the comments section. In so doing, I consider what I call the 'affective affordances' of the videos, including the main discourses, images and practices they depict in the attempt to generate affective forces. I also consider the gendered nature of the videos, which my analysis found was closely related to their affective affordances. I discuss these materials in more detail following a discussion of YouTube and food-related content.

The approach I adopted to analyse the YouTube videos discussed in this chapter was developed from my reading of feminist new materialism theory, and particularly the work of Jane Bennett (2001, 2010), Karen Barad (2003, 2014), Donna Haraway (Franklin & Haraway, 2017; Haraway, 2016) and Rosi Braidotti (2016, 2019). These scholars all view humans as inextricably part of more-than-human assemblages. They emphasise the lively agencies that are created by and distributed between humans and non-humans, including media artefacts such as digital videos and organic objects such as edible matter. Feminist new materialism also highlights the role of affective forces and relational connections as they are established via what Barad (2007) titles the 'intra-actions' of humans and non-humans when they come together and form dynamic assemblages. From this perspective, digital media offer ways of expressing and sharing affect between humans and non-humans.

Drawing on this scholarship, I argue that what Bennett (2009) describes as the 'thing-power' of these assemblages generates capacities and forces. The affordances of digital technologies such as the YouTube platform and the digital cameras used to create the videos posted on the platform come together with the affordances of human bodies (their capacities for movement, growth, sensory engagement, memory and the expression of feeling) in generative ways. My use of the terms 'affective force' and 'agential capacities' draws on the Spinozian/ Deleuzean interpretation of affects as embodied, relational, dynamic and shared intensities that have effects on the human and non-human actors experiencing them, including opening up capacities and connections (or in some cases, closing them off) (see, for example, Bennett, 2009). Adopting a more-than-human perspective, affect is understood as moving between humans and non-humans, generating responses and capacities. The forces of affects can have effects on micropolitical and macropolitical levels.

In digital media, humans and non-humans come together to generate affects in response to food-related topics and issues such as food production and processing, eating and cooking practices, human embodiment, gender relations, racial or ethnic politics, religious belief, climate change and animal rights (Forchtner & Tominc, 2017; Lavis, 2017; Lupton, 2017, 2019). As Haraway (Franklin & Haraway, 2017) describes it, humans are part of the 'compost' of materials with which they assemble as they move through their lives. These materials include other organic entities, which humans may consume as food or which may live in their digestive system as microflora, assisting with gut function and digestion. They have tangible effects on human bodies (Bennett, 2009). They can be literally incorporated into human flesh, unless excreted by the body, and move as dynamic flows into and out of human bodies into the more-than-human environment as part of movements of composition, re-composition and de-composition.

Food on YouTube

YouTube was launched in mid-2005 with the explicit intention of providing a platform for people to share home-made videos, or to 'broadcast themselves' (Arthurs, Drakopoulou, & Gandini, 2018). The platform has rapidly grown in popularity in recent years. A Pew Research Center survey found in 2018 that YouTube was used by nearly three-quarters of the adult respondents in the USA, including 94 per cent of respondents aged 18 to 24 years (Smith & Anderson, 2018). Pew has also found that YouTube has become the most popular online platform among American adolescents, well ahead of Instagram, Snapchat and Facebook (Anderson & Jiang, 2018).

A participatory sharing ethos was central to YouTube from the beginning, with an emphasis on broadcasters revealing intimate details of their lives, often by making videos of themselves talking directly and frankly to camera (Arthurs et al., 2018; Marwick, 2015; Raun, 2018). This has allowed members of marginalised or stigmatised social groups such as LGBTQI people (Lovelock, 2017; Raun, 2018) to receive a public voice and develop like-minded communities using vlogging (video blogging). YouTube therefore offers a potentially vast and diverse range of lifestyles, political views and identities to receive expression and publicity, including controversial and contentious viewpoints. Over time, the participatory ethos promoted by YouTube has become commercialised, with 'influencers' and 'micro-celebrities' (Hou, 2018; Khamis, Ang, & Welling, 2017; Marwick, 2015; Raun, 2018) taking advantage of the potentially huge audiences available to them to compete for their attention. YouTubers can commoditise their channels based on their popularity by being paid by Alphabet (the company that own both Google and YouTube) to display advertisements in their videos or by attracting companies that will sponsor their content or pay to have their products recommended by the YouTuber. YouTubers are now expected to create a personal channel accompanied by a personal brand as part of commoditising their content, while simultaneously maintaining intimacy and authenticity (Arthurs et al., 2018). Accumulating subscribers to their channel is a particularly sought-after activity for aspiring micro-celebrities on YouTube (García-Rapp, 2017).

The rise of professional YouTubers has coincided with the popularity of food-related content on the platform. Google's analysis of food videos on YouTube in 2014 (Delgrado et al., 2014) noted that food content was rapidly increasing in volume and popularity: that year alone, there was a 280 per cent growth in food channel subscription, views of food and recipe content increased by 59 per cent and social engagement such as likes, shares and comments rose by 118 per cent. A survey conducted by Google found that nearly half of all adults watch YouTube food videos, while younger adults (aged 18 to 34) viewed 30 per cent more food content than other adult age groups. Both men and women in this age group were keen food video viewers, particularly to learn cooking techniques and try new recipes. In another analysis, Google researchers noted that 'how-to' food videos are very popular on YouTube, with 'how to cook ...' one of the ten most popular searches on the platform (Cooper, 2015). Popular 'FoodTubers' include celebrity chefs who are well known from other media appearances, such as Jamie Oliver, but also amateur cooks who have hit on a successful formula in their 'how-to-cook' video presentations (Goodman et al., 2017).

Despite the popularity of food-related YouTube videos, few cultural analyses of this genre of digital media have been conducted. One example is an analysis by Lavis (2015), examining YouTube videos posted by women identifying as BBW (big beautiful women), some of which showed them consuming food. These videos directly bring together the sensuality of unrestrained food consumption with that of abundant female flesh. These videos are unusual in depicting fat women who openly and freely eat whatever they like, in direct contravention to proscriptions on fat women eating the 'wrong' foods in public (Pausé, 2015). Another study focused on competitive food eaters, who often use social media and YouTube to publicise their efforts. It found that these competitors use these platforms to present themselves as highly trained and disciplined 'ingestors'/athletes rather than as grotesque and disgusting freaks, as popular media narratives often portray them (Abbots & Attala, 2017). A third study investigated the subculture of food-related YouTube videos made by 'Balaclava Kuche [Kitchen]', a German neo-Nazi group comprising several men and one woman who post vegan cooking videos that serve as a direct counter to assumptions that veganism is limited to peace-loving, left-leaning women. The videos show the team preparing vegan food wearing balaclavas and t-shirts with Nazi slogans, engaging in banter using coded references to the support of Nazi philosophies (Forchtner & Tominc, 2017).

Stephanie Buttermore's cheat day videos

At first glance, Stephanie Buttermore's and the Epic Meal Time team's video channels appear to be very different. The videos made by Buttermore for her channel – entitled 'Stephanie Buttermore, Ph.D.' (Stephanie Buttermore Ph.D., 2019) – mostly subscribe to the 'fitspo' (short for 'fitness inspiration') and 'clean eating' genres of online content. The channel banner features a glamour colour image of Buttermore in skimpy workout gear, accompanied by a black-and-white close-up image of her in a laboratory setting, peering down a microscope, to signify her scientific training and credentials. At the end of 2019, Buttermore's channel had 814,000 subscribers. Most of her videos provide advice about nutrition and workouts to build the slim and strong, yet glamorous and normatively feminine body that is idealised in fitspo cultures (Lupton, 2017, 2018b, 2019; Riley & Evans, 2018). Her videos combine workout walk-throughs, reviews of the science of health and fitness and tips with nutrition advice for better body-building and health promotion, as well as the occasional make-up tips video. As such, Buttermore's videos conform to the entrepreneurial femininity that is promoted across the internet by young women eager to engage in personal branding and gain influencer status (Abidin, 2016; Abidin & Gwynne, 2017; Khamis et al., 2017).

The cheat day videos that Buttermore makes, however, are in sharp contrast from the other content on her channel. They depict her spending an entire day driving around to various fast food outlets, bakeries and supermarkets purchasing and consuming large quantities of the high-fat, high-sugar and highly processed foods that are usually proscribed by fitspo and clean eating advocates. The cheat day videos comprise approximately one-fifth of Buttermore's output, averaging approximately one video a month. These videos are her most popular YouTube output: some have received over one million views, far more than her videos focusing on health and fitness, which tend to attract a few hundred thousand views.

The cheat day videos typically begin with Buttermore speaking to camera early in the morning and describing her excitement about being able to consume all the food she desires throughout the day. One of her most highly viewed cheat day videos (with over 3.4 million views and almost 5,000 comments by viewers by the end of 2018) is entitled 'I Ate Everything I Wanted for One Day ... (Fantasy Cheat Day)'. As the word 'fantasy' in this title suggests, Buttermore attempts to offer her followers the voyeuristic pleasure of her real-life extreme consumption, as well as the fantasy of over-consuming forbidden foods with no need for shame, guilt or accountability. There are similarities in this approach to the South Korean *meokbang* phenomenon, which also often involves young women live-streaming their practices of eating vast quantities of food using webcams. Some of these professional eaters also use YouTube to disseminate their eating feats (Donnar, 2017).

Some of the cheat day videos begin with Buttermore displaying her body in a bikini and conducting a weigh-in. She then goes on to film her food consumption for that day, accompanied by a calorie counter shown on the screen that reveals the accumulating number of calories she ingests by the day's end. The cheat day foods she consumes include typical 'bad' or 'unhealthy' high-calorie foods: doughnuts, pastries, fast food such as hamburgers, fries and pizza, chocolate bars, crisps, cookies, ice cream and so on. In between enthusiastic bites, Buttermore constantly talks to camera, detailing how much she is enjoying eating them. After she bites into one of these food items, she often shows the munched side in close-up to camera, so that viewers can see her perspective of the food as she consumes it. The close-ups of her highly madeup face and mouth as she daintily but enthusiastically consumes these foods draw on overtly sexualised imagery of hyper-feminine embodied pleasure that is familiar from commercial advertising. Buttermore's narrative during the cheat day videos constantly emphasises that she feels no guilt at all about eating these foods as she considers cheat days to be a useful adjunct to her otherwise highly disciplined lifestyle. The everyday surroundings and constant chatter provided by Buttermore as she purchases and consumes these products also conform to the norms of YouTubing as an intimate and confessional insight into presenters' everyday thoughts, feelings and behaviours (Hou, 2018).

Like the competitive eaters who post videos of their feats on YouTube, Buttermore's videos celebrate her apparent ability to 'gurgitate' (ingest and contain) (Abbots & Attala, 2017) a higher volume of food than her usual intake without apparent ill effect. Indeed, there is an element of competition in Buttermore's videos, as she shows the next day how little her binge eating has affected her body size and weight. Buttermore's cheat day videos include appraisals at the end of the day or the next morning in which she shows off her body and weighs herself and comments on how many pounds she has put on as a result of her bingeing. Despite a slight weight gain (usually a few pounds) and slightly more round shape on an otherwise very flat and muscular abdomen, her body looks hardly different from the morning before she ate all the forbidden food depicted in the video.

Buttermore's 'before-and-after' format brings the narrative of her cheat day to a satisfying narrative of indulgence devoid of punishment. Just as the competitive eaters' self-representation focuses on their high-levels of self-control and disciplined eating habits to become successful at the food challenges in which they compete (Abbots & Attala, 2017), Buttermore's binge eating is similarly controlled for maximum 'wow' factor for her audiences and minimum (apparent) long-term effects on her otherwise fitspo-ideal body. As a result, of her highly controlled and aestheticised mode of excessive eating, her body never transgresses to the point of inspiring abjection or disgust. The comments made by viewers to her YouTube videos frequently emphasise the vicarious pleasure they gain from watching Buttermore consume these kinds of foods with impunity. Commentators often note their disbelief that such a slight, fit-looking woman can consume so much food, and how fortunate she is to be able to indulge herself without needing to worry about consequent effects on her health, body weight or fitness levels, and therefore be free from the guilt, shame or self-disgust that others would feel after such binges. Comments such as 'How can she eat that much OMG I would eat two donuts and be full', 'Watching you eat is satisfying' and 'I like watching these videos just so I can look at all the food' are common.

Ultimately, therefore, Buttermore is able to indulge herself but continue to triumph in maintaining her idealised fitspo body, in what might be the ultimate fantasy of combining self-discipline with self-indulgence. Viewers of her videos can share her pleasure at relinquishing her tight control over her body, safe in the knowledge that this abandonment of self-discipline has no ill effects on her appearance or health.

Epic Meal Time videos

Epic Meal Time is one of the most well-known and popular of the genre of excessive food on YouTube. At the time of writing (late 2019), the channel had attracted over seven million subscribers (this can be compared with celebrity chefs Jamie Oliver's four million and Gordon Ramsay's almost 12 million subscribers). The Epic Meal Time team had uploaded over 400 videos to their channel, all following a standard approach of showing the men bantering about how much they love meat and fast food and then creating a fantasy meal in which vast quantities of meat and fast food (and occasionally, sweet dessert-style dishes) are put together using minimal preparation.

The emphasis of the Epic Meal Time videos is on combining high-calorie and junk foods in 'epic' ways that exceed any previous attempts. All the food stuffs used are deliberately employed because of their cultural status as contaminating of the body due to their high-fat, high-sugar and highly processed qualities. These meals are positioned in stark counterpoint to diets such as those featuring 'clean', 'sustainable' or 'healthy' eating. Each video ends with members of the team biting into the food they have created with the appearance of hungry men desperate to consume it – although tellingly, they are typically shown only taking a few mouthfuls rather than eating significant quantities.

Like Buttermore's cheat day videos, quantifying the calorie content of the foods consumed during the video as they accumulate is part of the imagery. Thus, for example, the 'World Record Most Bacon BLT' features hundreds of bacon rashers with mayonnaise made from bacon oil and bacon bits, while the title of the 'Over 70 Big Macs in this Mac & Cheese' involves wrapping multiple Big Macs in pasta dough, cooking several of these items and then heaping them in a huge bowl and throwing melted cheese and bacon over them to create a 'Big Mac Big Mac'. Many meals consist of strange, incongruous concoctions, such as cheeseburger ramen. The trope of accumulation, indeed, is dominant in both Buttermore's channel and Epic Meal Time, with commentary constantly referring to 'more', titles describing 'giant', 'world record' and 'supersize' meals, the calorie count multiplying and the burgeoning volume of food. This is 'freak' food.

The Epic Meal Time videos draw on key elements of 'bro culture' (in the UK often referred to as 'lad culture', and in the USA also related to 'frat' or 'dude' culture), a form of hypermasculinity that relies on representing men as fun-loving, emotionally immature and instinctual creatures who are mainly focused on the pleasures of indulging their desires (Nichols, 2018). Bro culture is often sexist and misogynistic, reproducing aggressive orientations towards women that depict them as purely objects for the indulgence of men's lust. It celebrates excesses in alcohol, food and porn consumption. Everyday forms of sexism in bro culture typically use humour and banter as ways of denigrating and objectifying women (Nichols, 2018). The comments on Epic Meal Time are overwhelmingly from men, who usually engage in bantering and admiring tones, making reference to how much they love the ingredients or concept, how much vegans or vegetarians would be repulsed by the content or how 'artery-clogging' the food is.

The Epic Meal Time videos involve liberal swearing and sexualised references to food – particularly meat and parts of animals such as cow 'butts', which are frequently compared to women's buttocks (Lupton, 2019). These videos draw on long-held cultural tropes that equate the desire for and consumption of red meat with hypermasculinity, and women as meat-like objects (Adams, 2010; Lupton, 2019). The men's longing to consume meat is routinely portrayed as akin to sexual desire and even romanticised. As the opening commentary in the 'A Date with Bacon' video (subtitled 'Bacon's My Bottom Bitch') put it (addressing a strip of bacon): 'I want to be all over you tonight – I want you to be all over me.' Viewers are encouraged to 'thumbs up and favourite this video because you would marry bacon if it were socially acceptable'. This and other bacon-focused videos on Epic Food Time serve to highlight and play with the dominant meanings of bacon in North American mainstream culture as a transgressive yet enticing and particularly irresistible food, as pointed out by Weiss (2018).

Discussion

In the carnivalesque food videos discussed in this chapter, a trajectory of historical Western concerns about food can be traced, along with newer preoccupations and moral meanings. The struggle between disciplining the body by controlling the volume and type of food that is consumed and giving into the desires of the flesh can be identified from popular media texts as far back as the early modern period (Appelbaum, 2008). When it comes to food excesses, there is a very fine line between pleasure and disgust (Miller, 1997). According to Thompson (2007, p. 114), 'The carnivalesque body is equally copulative and excretory, salacious and scatological, vibrant and on the way to becoming decomposing humus that feeds the earth.' In accounts such as the videos examined in this chapter, the more-than-human qualities of the food–human assemblage and the affects that are distributed with and through these assemblages become particularly evident: Haraway's compost metaphor is particularly apposite.

The boundary between pleasure and disgust is never stable, as it is dependent on people's embodied life histories, acculturation and broader historical and cultural contexts. The YouTube videos I have referred to in this chapter have been created in a cultural environment in which disgust at fat bodies has intensified (Lupton, 2018b), normative embodied modes of femininity privilege eating 'healthy' foods for both good health and physical attractiveness (Cairns & Johnston, 2015; Lupton, 2018b), over-consumption of food is considered uncivilised, irresponsible and distasteful (Abbots & Attala, 2017; Lupton, 2018b; Phillipov, 2013) and the worship of the tightly disciplined fitspo body for both women and men is pronounced (Deighton-Smith & Bell, 2018; Lupton, 2017). At the same time, anxieties about 'contaminated' food have given rise to movements such as clean eating and a growing awareness of and sensitivity to animal suffering and environmental degradation has propelled activist movements and personal dietary practices such as organic food consumption, vegetarianism and veganism (Schneider, Eli, Dolan, & Ulijaszek, 2018). These cultural currents are operating in a technological environment in which it is easier than ever for people to produce and disseminate their own videos and attract large audiences, becoming internet 'influencers' who in some cases can profit from their popularity.

As I have shown, Buttermore's cheat day and the Epic Meal Time videos intra-act with audiences to reproduce sociocultural discourses, meanings and practices concerning affect, food consumption, excess, embodiment and gender norms. Elsewhere (Lupton, 2018b, 2019) I have discussed how digital media artefacts such as GIFs and memes on food-related topics often refer to the struggle people have over restraining their desire for foods that are culturally coded as forbidden or contaminating to bodily integrity. The videos made by Buttermore and the Epic Meal Time team similarly generate affective forces from their combinations of words (how the video-makers describe what they are doing) and images. The meanings of the videos and their agential capacities – their thing-power – are built on the existing imaginaries adhering to types of food, how they should be combined, and how they should be eaten. Their qualities as transgressive, grotesque and disturbing generate a fascination that explains their popularity. These videos, therefore, share some similarities with other digital media representations of excessive food consumption, including portrayals of competitive eaters (Abbots & Attala, 2017) and fat women showing themselves eating large quantities of food for the sexual gratification of their followers (Lavis, 2015; Woolley, 2017).

In both the Epic Meal Time oeuvre and Buttermore's videos, a calorie counter is used to demonstrate in quantified/scientific terms the indulgence of the food consumed. Part of the pleasure of these indulgences is in flouting the disciplinary norms of calorie-counting. In these videos, the higher the calories, the better – the greater the achievement – in stark contradistinction with the limits on calorie consumption that weight-loss or 'healthy' diets demand. Just as viewers can watch Buttermore and the Epic Meal Time men eat their way through the high calorie foods, they can monitor the ever-rising calorie counter in a corner of the screen that acts as a symbol of the quantities of food that is entering their bodies. This responsive calorie counter is the latest in a long line of metricising food intake and using the numbers so generated to convey moral assumptions about the worth and quality of food and the bodies that ingest it (Crawford, Lingel, & Karppi, 2015). In the case of these carnivalesque videos, the calorie counter has a different meaning: it is a way of quantifying, demonstrating and celebrating excess.

The examples of Buttermore's cheat day videos and Epic Meal Time's grotesque cooking videos bring to the fore the gendered dimensions of carnivalesque consumption in these media. Gender norms are central to assumptions about the appropriateness of certain foods and food quantities that should be consumed. Normative idealised femininities tend to be associated with a diet of salads and other vegetables, 'dainty' eating and highly controlled eating or dieting for the sake of health and a slim body (Cairns & Johnston, 2015; Lupton, 2017, 2018b), while dominant ideals of masculinity assume that men favour red meat and fast food and need large quantities of food, exhibiting a cavalier disregard for health (Lupton, 1996; Phillipov, 2013).

In some ways, these carnivalesque food videos bear similarities with other digital media characterised as 'food porn', in which the multisensory delights and affective forces of food are emphasised (Lavis, 2017; Rousseau, 2013; Taylor & Keating, 2018). The use of the word 'porn' points to the common associations between the carnal enjoyment of sexual pleasure and that relating to food consumption. The types of food depicted in food porn may be beautiful to look at, celebrated for their 'healthy' or 'clean' qualities or represented as an expensive indulgence, but food porn can also include foods that are culturally designed as 'naughty', 'bad' or 'forbidden' (Lupton, 2019; Rousseau, 2013). Food porn media, therefore, can evoke the visceral responses of both pleasure and disgust (including self-disgust for lusting after the food depicted). This is a particularly difficult boundary for women to negotiate, given the negative associations of self-indulgence in food and lack of self-discipline that tend to be represented in dominant representations as archetypally feminine and the intensified focus on the importance of the contained, slim female body (Dejmanee, 2016; Lupton, 2018b).

One major departure from typical food porn videos exhibited in the videos analysed in this chapter, however, is their transgressive and excessive nature. If the Epic Meal Time videos are a kind of 'food porn', they are akin to the 'gonzo porn' genre: visual representations of sexual acts that are deliberately grungy – to the point of ugliness – departing from aestheticised norms of sexualised embodiment (Saunders, 2018). Gonzo porn images rely for their appeal not on the beauty of the objects depicted, but rather their transgressive nature that disrupts and challenges conventional norms of how these things should look. This porn genre is interested in inspiring visceral and ambivalent reactions in audiences, with an emphasis on intense corporeality in images that may generate disgust just as much as desire (Saunders, 2018). Buttermore's cheat day videos are far more aestheticised than the Epic Meal Time corpus, but their overt depiction of extreme indulgence is their most shocking feature, given their position within the Buttermore's oeuvre of fitspo messaging. It is Buttermore's unashamed greed for and consumption of 'bad' food rather than the visual aesthetics of her videos that accord most closely to gonzo porn, which in some cases can celebrate and emphasise women's sexual desires and pleasures that may depart from normative assumptions about what women want (Saunders, 2018).

The grotesque and transgressive nature of Buttermore's cheat videos reside in the stark contrast of her food consumption with her otherwise highly controlled and contained appearance. The fact that she depicts herself consuming (and enjoying) junk and sugary food all day long is part of the transgressive pleasure. She allows herself not only one or two treats, but many, which accumulate to a much greater high-calorie total for the day than her usual diet allows for. Buttermore is deliberately contaminating her body by eschewing her usual 'clean' and healthy eating habits. For the Epic Meal Time bros, the food they present accords with their dude presentations, but takes it to extremes as a form of entertainment. They may well consume junk foods regularly as part of their normal diets, but their 'epic' meals go well beyond this usual consumption. For the Epic Meal Time bros presenting their outrageous recipes and messily consuming the results with gusto, a version of hyper-masculinity is presented that includes elements of misogyny and the objectification of women. Meat-eating as a masculine practice, in particular, is emphasised in Epic Meal Time, but so too is cavalier disregard for health injunctions (Lupton, 2019).

To conclude, the expression and appreciation of carnivalesque food preparation or consumption in these videos offer a way for content creators and their audiences to express and celebrate their longing and desire to indulge in the fantasy of revelling in the sensory delights of forbidden food with no guilt or shame. Simultaneously, however, these videos reproduce sexualised stereotypes of hyper-femininities and hyper-masculinities and surface forceful affective undercurrents of anxiety and ambivalence concerning excessive or wrong food consumption, revealing the fraught nature of contemporary digital food cultures.

Further reading

Contois, E.J.H. (2018). The spicy spectacular: Food, gender, and celebrity on Hot Ones. *Feminist Media Studies, 18*(4), 769–773.

Hart, D. (2018). Faux-meat and masculinity: The gendering of food on three vegan blogs. *Canadian Food Studies/La Revue canadienne des études sur l'alimentation, 5*(1), 133–155.

Turner, B. (2019). *Taste, Waste and the New Materiality of Food*. London: Routledge.

References

Abbots, E.-J., & Attala, L. (2017). It's not what you eat but how and that you eat: Social media, counter-discourses and disciplined ingestion among amateur competitive eaters. *Geoforum, 84*, 188–197.

Abidin, C. (2016). Visibility labour: Engaging with Influencers' fashion brands and #OOTD advertorial campaigns on Instagram. *Media International Australia, 161*(1), 86–100.

Abidin, C., & Gwynne, J. (2017). Entrepreneurial selves, feminine corporeality and lifestyle blogging in Singapore. *Asian Journa of Social Science, 45*(4–5), 385–408.

Adams, C.J. (2010). Why feminist-vegan now? *Feminism & Psychology, 20*(3), 302–317.

Anderson, M., & Jiang, J. (2018). *Teens, social media and technology 2018*. Retrieved from www.pewinternet.org/2018/05/31/teens-social-media-technology-2018/.

Appelbaum, R. (2008). *Aguecheek's Beef, Belch's Hiccup, and Other Gastronomic Interjections: Literature, Culture, and Food among the Early Moderns.* Chicago: University of Chicago Press.

Arthurs, J., Drakopoulou, S., & Gandini, A. (2018). Researching YouTube. *Convergence, 24*(1), 3–15.

Bakhtin, M. (1984). *Rabelais and His World.* Bloomington: Indiana University Press.

Barad, K. (2003). Posthumanist performativity: Toward an understanding of how matter comes to matter. *Signs, 28*(3), 801–831.

Barad, K. (2007). *Meeting the Universe Halfway: Quantum Physics and the Entanglement of Matter and Meaning.* Durham: Duke University Press.

Barad, K. (2014). Diffracting diffraction: Cutting together-apart. *Parallax, 20*(3), 168–187.

Bennett, J. (2001). *The Enchantment of Modern Life: Attachments, Crossings, and Ethics.* Princeton: Princeton University Press.

Bennett, J. (2009). *Vibrant Matter: A Political Ecology of Things.* Durham: Duke University Press.

Bennett, J. (2010). A vitalist stopover on the way to a new materialism. In D. Coole & S. Frost (Eds.), *New Materialisms: Ontology, Agency and Politics* (pp. 47–69). Durham, NC: Duke University Press.

Braidotti, R. (2016). Posthuman critical theory. In D. Banerji & M. Paranjape (Eds.), *Critical Posthumanism and Planetary Futures* (pp. 13–32). Berlin: Springer.

Braidotti, R. (2019). A theoretical framework for the critical posthumanities. *Theory, Culture & Society, 36*(6), 31–61.

Cairns, K., & Johnston, J. (2015). Choosing health: Embodied neoliberalism, postfeminism, and the 'do-diet'. *Theory and Society, 44*(2), 153–175.

Cairns, K., Johnston, J., & Baumann, S. (2010). Caring about food: Doing gender in the foodie kitchen. *Gender & Society, 24*(5), 591–615.

Cooper, J. (2015). Cooking trends among millennials: Welcome to the digital kitchen. Retrieved from www.thinkwithgoogle.com/consumer-insights/cooking-trends-among-millennials/.

Crawford, K., Lingel, J., & Karppi, T. (2015). Our metrics, ourselves: A hundred years of self-tracking from the weight scale to the wrist wearable device. *European Journal of Cultural studies, 18*(4–5), 479–496.

Cronin, J.M., McCarthy, M., & Collins, A. (2014). Creeping edgework: Carnivalesque consumption and the social experience of health risk. *Sociology of Health & Illness, 36*(8), 1125–1140.

De Solier, I. (2018). Tasting the digital: New food media. In K. LeBesco & P. Naccarato (Eds.), *The Bloomsbury Handbook of Food and Popular Culture* (pp. 54–65). London: Bloomsbury.

Deighton-Smith, N., & Bell, B.T. (2018). Objectifying fitness: A content and thematic analysis of #fitspiration images on social media. *Psychology of Popular Media Culture, 7*(4), 467–483.

Dejmanee, T. (2016). 'Food porn' as postfeminist play: Digital femininity and the female body on food blogs. *Television & New Media, 17*(5), 429–448.

Delgrado, J., Johnsmeyer, B., & Balanovskiy, S. (2014). Millennials eat up YouTube food videos. Retrieved from www.thinkwithgoogle.com/consumer-insights/millennials-eat-up-youtube-food-videos/.

Donnar, G. (2017). 'Food porn' or intimate sociality: Committed celebrity and cultural performances of overeating in *meokbang*. *Celebrity Studies, 8*(1), 122–127.

Forchtner, B., & Tominc, A. (2017). Kalashnikov and cooking-spoon: Neo-Nazism, veganism and a lifestyle cooking show on YouTube. *Food, Culture & Society, 20*(3), 415–441.

Franklin, S., & Haraway, D. (2017). Staying with the manifesto: An interview with Donna Haraway. *Theory, Culture & Society, 34*(4), 49–63.

García-Rapp, F. (2017). Popularity markers on YouTube's attention economy: The case of Bubzbeauty. *Celebrity Studies, 8*(2), 228–245.

Goodman, M.K., Johnston, J., & Cairns, K. (2017). Food, media and space: The mediated biopolitics of eating. *Geoforum, 84*, 161–168.

Haraway, D. (2016). *Staying with the Trouble: Making Kin in the Chthulucene.* Durham: Duke University Press.

Hou, M. (2019). Social media celebrity and the institutionalization of YouTube. *Convergence, 25*(3), 534–553.

Khamis, S., Ang, L., & Welling, R. (2017). Self-branding, 'micro-celebrity' and the rise of Social Media Influencers. *Celebrity Studies, 8*(2), 191–208.

Lavis, A. (2015). Consuming (through) the Other? Rethinking fat and eating in BBW videos online. *M/C Journal, 18*(3). Retrieved from http://journal.media-culture.org.au/index.php/mcjournal/article/view/973.

Lavis, A. (2017). Food porn, pro-anorexia and the viscerality of virtual affect: Exploring eating in cyberspace. *Geoforum, 84*, 198–205.

Lewis, T. (2018). Digital food: From paddock to platform. *Communication Research Practice, 4*(3), 212–228.

Lovelock, M. (2017). 'Is every YouTuber going to make a coming out video eventually?': YouTube celebrity video bloggers and lesbian and gay identity. *Celebrity Studies, 8*(1), 87–103.

Lupton, D. (1996). *Food, the Body and the Self.* London: Sage.

Lupton, D. (2017). Digital media and body weight, shape, and size: An introduction and review. *Fat Studies, 6*(2), 119–134.

Lupton, D. (2018a). Cooking, eating, uploading: Digital food cultures. In K. LeBesco & P. Naccarato (Eds.), *The Handbook of Food and Popular Culture* (pp. 66–79). London: Bloomsbury.

Lupton, D. (2018b). *Fat* (2nd ed.). London: Routledge.

Lupton, D. (2019). Vitalities and visceralities: Alternative body/food politics in new digital media. In M. Phillipov & K. Kirkwood (Eds.), *Alternative Food Politics: From the Margins to the Mainstream* (pp. 151–168). London: Routledge.

Marwick, A.E. (2015). You may know me from YouTube. In P.D. Marshall & S. Redmond (Eds.), *A Companion to Celebrity* (pp. 333–349). Chicester: John Wiley.

Miller, W. (1997). *The Anatomy of Disgust.* Cambridge, MA: Harvard University Press.

Nichols, K. (2018). Moving beyond ideas of laddism: Conceptualising 'mischievous masculinities' as a new way of understanding everyday sexism and gender relations. *Journal of Gender Studies, 27*(1), 73–85.

Pausé, C. (2015). Rebel heart: Performing fatness wrong online. *M/C Journal, 18*(3). Retrieved from www.journal.media-culture.org.au/index.php/mcjournal/article/view/977.

Phillipov, M. (2013). Resisting health: Extreme food and the culinary abject. *Critical Studies in Media Communication, 30*(5), 377–390.

Probyn, E. (2003). *Carnal Appetites: FoodSexIdentities.* London: Routledge.

Raun, T. (2018). Capitalizing intimacy: New subcultural forms of micro-celebrity strategies and affective labour on YouTube. *Convergence, 24*(1), 99–113.

Riley, S., & Evans, A. (2018). Lean light fit and tight: Fitblr blogs and the postfeminist transformation imperative. In K. Toffoletti, H. Thorpe, & J. Francombe-Webb (Eds.), *New Sporting Femininities: Embodied Politics in Postfeminist Times* (pp. 207–229). Cham: Palgrave Macmillan.

Rousseau, S. (2012). *Food and Social Media: You Are What You Tweet*. Lanham, MD: Rowman Altamira.

Rousseau, S. (2013). Food 'porn' in media. In D. M. Kaplan (Ed.), *Encyclopedia of Food and Agricultural Ethics* (pp. 748–754). Dordrecht: Springer. Retrieved from https://link.springer.com/referenceworkentry/10.1007%2F978–94–007–6167–4_395–1.

Saunders, R. (2018). Grey, gonzo and the grotesque: The legacy of porn star Sasha Grey. *Porn Studies, 5*(4), 363–379.

Schneider, T., Eli, K., Dolan, C., & Ulijaszek, S. (Eds.). (2018). *Digital Food Activism*. London: Routledge.

Smith, A., & Anderson, M. (2018). Social media use in 2018. Retrieved from www.pewinternet.org/2018/03/01/social-media-use-in-2018/.

Stephanie Buttermore Ph.D. (2019). [Video channel] Retrieved from www.youtube.com/channel/UC4gDYbCEIb69uvFmCX5Lyuw

Taylor, N., & Keating, M. (2018). Contemporary food imagery: Food porn and other visual trends. *Communication Research and Practice, 4*(3), 307–323.

Thompson, C.J. (2007). A carnivalesque approach to the politics of consumption (or) grotesque realism and the analytics of the excretory economy. *The Annals of the American Academy of Political and Social Science, 611*(1), 112–125.

Weiss, B. (2018). 'Life-changing bacon': Transgression as desire in contemporary American tastes. *Food, Culture & Society, 21*(5), 664–679.

Woolley, D. (2017). Aberrant consumers: Selfies and fat admiration websites. *Fat Studies, 6*(2), 206–222.

PART 2
Healthism and spirituality

4

YOU ARE WHAT YOU INSTAGRAM

Clean eating and the symbolic representation of food

Stephanie Alice Baker and Michael James Walsh

Introduction

Digital technologies have altered the way that many people consume food. Whereas food was traditionally consumed in co-present situations, digital technologies function as 'disembedding mechanisms' that 'lift out' social relations from local contexts of interaction so they can be experienced across indefinite spans of time-space (Giddens, 1992, pp. 21–22). The result is a profoundly different understanding of food and its relationship to physical space. While the internet allows information to be communicated at an unprecedented rate, social media facilitate social interaction among online communities. Social media sites, such as Instagram, alter how we treat public space. Free from the confines of co-presence, hashtags can be used on these platforms to access like-minded communities at any time and from any space (Baker & Walsh, 2018).

The ubiquity of digital technologies has transformed how many people consume food, and the meanings and discourses they ascribe to it. People no longer rely solely on experts to provide information about what food to eat. Instead, they turn to those with influence on the internet to give advice (Baker & Rojek, 2019a). Influencers typically use blogs and social media platforms to document their lives and lifestyles, and to market products and services for social and economic gain (Page, 2012). They have become a common feature of contemporary culture due to a series of technological developments. The prevalence of social media, mobile broadband and digital devices enable users to transmit their message to a global audience. The ubiquity of camera phones allows meaning to be communicated visually in photo and video form. There is a credibility associated with the image with much of the influence of online communication generated through aesthetically attractive photos and videos (Baker, 2014; Baker & Rojek, 2019a, 2019b). These images, and the messages they communicate, reveal much about the social and cultural meanings of food and eating practices. They convey not only the discourses behind food movements, but also how food and the body are used to symbolise the moral character and identity of the user.

Clean eating is a popular dietary trend that has achieved widespread public attention in the last decade. While there is no single definition of the term, clean eating generally refers to the consumption of unprocessed food considered to be as close to its whole form and natural state as possible (McCartney, 2016). Those foods perceived to be 'clean' and 'pure' are often contrasted with food that has undergone greater processing and refining and has been treated with pesticides or combined with additives (Nevin & Vartanian, 2017). The clean eating movement is defined *in relation* to what it is not, with practitioners restricting certain food groups perceived to be 'unclean' and 'impure'. Sugar, gluten, dairy and red meat are the foods most commonly abstained from, in addition to chemicals, additives and preservatives (Allen, Dickinson, & Prichard, 2018). The omission of these food groups is largely a reaction to the food system and its perceived failure (Goodman, DuPuis, & Goodman, 2012). Some clean eating advocates are purely plant-based, while others eat fish or meat.

In this chapter, we explore how 'clean eating' communities present themselves online. We perform visual content analysis of food images on Instagram to examine the social and cultural meanings of clean eating and food. Instagram is a social network app designed to share photos and videos from a mobile smartphone or device. It is one of the fastest growing online social networks, especially among people aged 18 to 25 years (Smith & Anderson, 2018). Drawing upon and developing cultural approaches to social interaction, we employ the concept of the 'affirmation ritual' to understand how status and identity are collectively established online.

Food and moral emotions: purity, disgust and contamination

While diverse diets label themselves as 'clean' (e.g., vegan, ketogenic or paleo), the clean eating movement is fuelled by a common narrative which constructs modern farming and technology as artificial and impure in relation to an idealised past when food production and consumption was simpler and less refined (Baker & Rojek, 2019a). Clean eating is described as a lifestyle rather than a diet. It is presented as a way of living to be adopted permanently instead of a short-term practice designed to achieve weight loss. As such, clean eating becomes central to identity with the term used to express the moral ethos of the consumer.

The relationship between food and morality has an established history. In many major religions, food is of profound theological significance. In Christianity, the apple is a symbol of temptation and original sin, bread and wine the means by which the believer exercises communion with Christ. Ritual food practices involving food preparation and consumption are fundamentally rooted in religious beliefs (Bailey, 2014). Muslims fast during Ramadan, while the dietary laws of Hinduism forbid consuming beef, and Judaism and Islam prohibit the consumption of pork. Prior to the professionalisation of the science of nutrition in the twentieth century, eating practices were largely defined by these religious beliefs; which provided a moral justification for what was considered a healthy diet (Mudry, 2018). Contemporary discourse around food is commonly infused with theological language. Food is described as heavenly, virtuous and indulgent, leading people into temptation and evoking feelings of guilt in its disciples when they indulge in sinful pleasure.

Emily Contois (2015) demonstrates how religiosity of this kind operates within dieting approaches that are generally considered to be secular. Scientists and medical experts appropriate the language of Christian Protestantism in order to construct a contemporary theology of weight loss. Diet manuals propagate the belief that weight loss requires conversion, sacrifice and commitment. This is achieved through rituals such as counting calories, tracking 'steps' and weighing oneself (Contois, 2015, p. 114). The language of diet literature treats thinness as a form of salvation (Lelwica, 2002) and 'fatness as a moral failure' (Contois, 2015, p. 122; Lupton, 2018b). Dieting theology assigns a moral value to food by constructing a dichotomy between 'good' and 'bad' foods and 'right' and 'wrong' ways of living. This moral logic postulates that to consume bad food is to be morally inferior, with good dietary choices culminating in the morally superior, thin individual. In this context the thin body operates as a signifier for health, self-control and moral virtue, so that you are what you eat both physically and metaphorically.

This binary moral logic underpins contemporary understandings of purity and dirt avoidance. In her seminal work, *Purity and Danger: An Analysis of Concepts of Pollution and Taboo* ([1966] 2002), Mary Douglas contends that ideas of purity and pollution are integral components of many cultures. She suggests that contemporary Western attitudes towards dirt and ideas about defilement differ from pre-modern cultures in two fundamental ways. First, dirt avoidance is perceived to be a matter of hygiene or aesthetics and not related to religion. Second, these understandings of dirt are informed by nineteenth-century discoveries of the bacterial transmission of disease and knowledge of pathogenic organisms. Douglas claims, however, that while our own conceptions of dirt are shaped by recent historical developments, the pollution beliefs that drive dirt avoidance are universal. She explains,

> If we can abstract pathogenicity and hygiene from our notion of dirt, we are left with the old definition of dirt as *matter out of place*. This is a very suggestive approach. It implies two conditions: a set of ordered relations and a contravention of that order. Dirt then, is never a unique, isolated event. Where there is dirt there is a system. Dirt is the by-product of a systemic ordering and classification of matter, in so far as ordering involves rejecting inappropriate elements. This idea of dirt takes us straight into the field of symbolism and promises a link-up with more obviously symbolic systems of purity.
>
> *(2002, p. 44 [emphasis added])*

From this perspective, purity is synonymous with social order, and subsequently an object (person or idea) is considered impure or dirty when it violates the social order as 'matter out of place'. Disgust is commonly felt towards those objects that threaten the social order with disgust sensitivity, and corresponding practices based around cleanliness, predicated on belief (Haidt, 2012). This Durkheimian view of social order emphasises the role of cultural beliefs and ritualised practices in establishing and maintaining symbolic boundaries and group identity.

The ritualised dimensions of identity and status

The concept of ritual is useful for understanding how status and identity are established online. Durkheim's later work on *The Elementary Forms of Religious Life* ([1912] 1995) explored the role of rituals in binding individuals to society. Late Durkheimian understandings of the religious dimensions of secular life demonstrate the continued role of rituals in creating symbols of group membership and energising individuals with emotional intensities (Alexander, 1990, 2004; Collins, 2004; Goffman, 1959). Erving Goffman (1967) notably applied Durkheim's approach to examine the role of rituals in upholding (and breaking) social organisation at the micro-level of social interaction. Alluding to the influence of the social situation over individual behaviour, Goffman noted that the object of sociological analysis was, 'not, then, men and their moments. Rather moments and their men' (1967, p. 3). However, Goffman's application of Durkheim was no replication; it posited considerable vulnerability in social structure, requiring individuals actively to restore social order through ritual interaction whenever disruptions arise (Burns, 1992, p. 31).

While Goffman's work examined standard face-to-face encounters prior to the advent of digital communication, his conceptual framework is useful for understanding self-presentation and status relations online (Hogan, 2010; Walsh & Clark, 2019). Social media are engineered to encourage users to seek affirmation and status as made measurable by metrics, such as likes, followers, reposts and comments (Marwick, 2013). The act of liking or commenting on a post represents a process of affirmation whereby an image is recognised and validated by others. When these metrics occur on those posts using a particular hashtag, they reveal more than status; they function as an online affirmation ritual – beyond the mere process of greeting described by Pate's (2006) 'acknowledgement ritual' – in which members of a particular community affirm one another's identities (Baker & Walsh, 2018). Online metrics operate as a means of symbolic exchange, signifying behaviours that both affirm validation and status.

Goffman's work on face-to-face encounters may initially appear unconnected with such concerns. However, his concept of ritual interchange is useful for considering the significance of how status is expressed online. As Brownlie and Shaw (2019, p. 104) have shown with respect to 'empathy rituals' on Twitter, Goffman provides a pertinent basis for understanding collective and aggregate expressions of emotions online through the analysis of banal interactions as opposed to more topical concerns (a strategy directly emulating Goffman's focus on seemingly prosaic face-to-face interaction).

Methodological approach

To examine the social and cultural meanings of clean eating and food, we performed visual content analysis of food images on Instagram. Similar to other social network sites, Instagram users create a profile and upload posts in the form of photos and videos to share with their followers. There is no technical requirement for reciprocity among followers. Because Instagram is public by default, anyone can view a user's posts unless they make their profile private. Instagram is configured around the contemporary moment, documenting images on a user's profile sequentially in the order in which they are uploaded. Posts can be produced, disseminated and consumed instantly from a user's smartphone, as denoted by the platform's prefix 'insta'. This configuration around the present gives the impression of spontaneity, despite that many images are carefully composed and edited before being posted and shared online.

Instagram is a predominantly visual medium. The application's interface renders imagery the primary mode of communication, providing users with a series of filters and editing tools (e.g., brightness, contrast, warmth) to enhance the quality and affects of their photos and videos. Visual imagery in the form of selfies (self-portrait photograph), photographs, videos, memes, GIFs and emojis are shared on the platform to create, attribute and share meanings with other users on the social network (Highfield & Leaver, 2016; Kaye, Malone, & Wall, 2017; Tiidenberg & Gómez Cruz, 2015; Walsh & Baker, 2017). Meaning is achieved through a series of practices including posting, captioning, editing, sharing, commenting, liking, reposting and deleting. Consequently, rather than viewing visual imagery as an isolated object, a more comprehensive way of viewing individual posts is to consider how they are positioned 'within an endless flow of posts, other people's posts, relationships, people, platforms, visual tropes and popular norms' (Tiidenberg, 2018).

Digital visual media are frequently used as part of political action or campaigns directed at resisting and reframing bodily norms (Lupton, 2018a), with individual posts contributing to a larger collective political narrative (Baker, 2014). However, despite the potential for user-generated content to challenge normative cultural values and conceptions of self, these visual representations can also reproduce hegemonic ideals pertaining to gender (Baker & Walsh, 2018; Döring, Reif, & Poeschl, 2016) and beauty (Abidin, 2014; Gill & Elias, 2014) with 'instafame' – visibility and attention on the platform – generally conferred upon conventionally attractive users (Marwick, 2015).

Although Instagram privileges images over words, one notable exception of popular text-based communication on the platform is hashtags. Users can add hashtags in the caption or comments section of their posts. If users add hashtags to a post that is set to public, the post will be visible on the corresponding hashtag page. A lifestyle blogger, for example, could upload an attractive picture of themselves in a bikini accompanied by the #cleaneating hashtag. The hashtag serves to categorise the post with other images using #cleaneating. Hashtags make content discoverable by a larger social network beyond a user's immediate followers, referred to by danah boyd (2011) as a 'networked public'. Hashtags also serve another important purpose: they bestow meaning to an image. In the example above, the hashtag makes a connection between clean eating and the attractive image, thereby attributing clean eating as the cause of the subject's physique (whether realised or not). Hashtags are deliberately added by users to ascribe meaning to their posts (Baker & Walsh, 2018). In this regard, hashtags provide a useful way to understand the symbols and discourses attributed to clean eating by the general public.

For our study, we examined photographs and videos categorised under the two most popular clean eating-related hashtags: #cleaneating and #eatclean. When searching for a hashtag on Instagram, the nine top posts and nine most recent posts are made visible at any point in time. 'Top posts' refer to 'trending hashtags and places' that depict 'some of the popular posts tagged with that hashtag or place' (Instagram, 2017). Trending in the context of social media refers to content or topics on a platform that are the most highly accessed or commented on. Instagram's application programming interface (API) ranks top posts according to the *quantity* of engagement (e.g., 'likes' and comments) and the *quality* (engagement rate of interaction by followers). 'Recent posts', conversely, privilege communication from the standpoint of the producer. This is because posts falling under this category are uploaded onto Instagram chronologically without considering a post's engagement rate or validation by other users. With the exception of 'likes' and comments, recent posts provide little sense of how a post has been received by the public. The value of analysing top posts for a given hashtag is that they signify those images validated by others (Baker & Walsh, 2018). Consequently, our focus on the representation of clean eating on Instagram pertains only to top posts – that is, those posts affirmed by users who viewed or 'engaged with' these hashtags.

Data were collected over an eight-day period in August 2017. The clean eating movement has an active presence on Instagram (at the time of data collection, between 30 and 40 million #eatclean and #cleaneating posts had been shared on the platform). Given that nine top posts were made visible each day for each of the two hashtags examined in this study, we collected 144 top posts in total using the #cleaneating and #eatclean. When collecting data, we captured screenshots of the posts including the image and other associated textual information (e.g., captions, comments and likes). Only public images were collected and analysed.

Once we collected the data, we analysed the posts to look for common characteristics among the images. Identifying common visual attributes, we inductively settled on eight codes to classify the posts. These are as follows:

1. *Glamour Shot* was coded if the person depicted was adorned in make-up or active wear.
2. *Kissy Face* was coded if the person depicted made this facial expression.
3. *Food* was coded if the focus of the photo was food.
4. *Before/After Shot* was coded if a collage was used to document a person's physical transformation over time.

5. *Muscle Presentation* was coded if the person posed to show off and provided an exaggerated display of their muscular physique.
6. *Advertisements* were coded if the post was used to advertise a service or product.
7. *Nature Shot* was coded if the image displayed a person in an outdoors natural setting.
8. *No Category* was used to describe those posts that did not clearly fit into one of the identifiable categories listed above.

Separate categories were also used to quantify the apparent gender of people depicted and the number of people portrayed in each image. In our discussion in this chapter, we focus on the first five categories listed above.

Findings

Health and gender

One of the most significant findings emerging from our sample was that the majority of clean eating top posts (76 per cent) did not include images of food. Those posts that did portray food using the #cleaneating and #eatclean hashtags generally fell into two categories. On the one hand, there were representations of fruit and vegetables in their 'natural', whole, unprocessed form. On the other hand, clean eating top posts featuring food included images of indulgent desserts. This was surprising, given that clean eating tends to involve the consumption of nutrient dense, low-calorie, protein-rich foods, such as green juices, vegetables, egg-white omelettes, fish and lean meat (Lupton, 2018a).

Although at first glance these images appeared to undermine clean eating principles by portraying high-caloric and sugary desserts, the hashtags that commonly accompanied these images indicated that the desserts were #healthy, #highprotein, #lowcarb, #weightloss, #vegan, #sugarfree and #crueltyfree, thereby, demonstrating that the desserts were not what Douglas (2002, p. 44) termed 'matter out of place' as they were free from the 'polluting' ingredients that drive dirt avoidance. In this context, the risk of transgressing the moral boundary of clean eating principles are negated by low-calorie substitutes (e.g., ice cream without cream, coffee without caffeine and gluten-free pasta) and ethical food practices (e.g., vegan desserts conveyed as a 'cruelty free' food practice). What is absent from the ingredients is vital to the food's symbolic significance and, by extension, the practitioner's moral character. The visual mimicry of indulgent desserts allows those who post these images to engage in normative ideals of food consumption that can become portrayed as morally and ethically defensible (Lupton, 2018a, p. 7).

The inclusion of indulgent desserts in clean eating hashtags also conforms to the trope of 'food porn'. The term is used to describe the enhancement of food-related images designed to excite a 'sense of the unattainable by proffering coloured photographs of various completed recipes' (Cockburn, 1977). In all cases, those foods represented using the clean eating hashtags were aesthetically enhanced using filters and editing techniques (e.g., cropping, brightness, colour and saturation) to appear more visually appealing. The representation of these foods on Instagram is not necessarily designed to be emulated by others. Rather, they are created to heighten the visual state of food and, by association, the user. In exaggerating the aesthetic qualities of food, these images draw on forms of idealisation that aim to garner esteem and attention to the subject and object of clean eating posts (Walsh & Baker, forthcoming).

Despite using hashtags that invoke acts of food consumption, most posts featured images of the body to signify the positive effects of clean eating. The prevalence of the healthy body was exemplified by the presence of three recurring visual themes: Kissy Face, Glamour Shot and the Before/After Shot. These visual tropes were deployed by users to signal status and self-improvement, specifically related to the subject's physical appearance. In posts of this kind, the body operates as a symbol to communicate and project an idealised version of the self to their networked audience. It is an example of what is termed 'aesthetic labour', the work involved in achieving conventional norms of attractiveness. This type of labour has an established history of disproportionately targeting women and informing their preoccupation with enhancing and regulating the body (Elias, Gill, & Scharff, 2017). Though relatively infrequent in our sample, those posts categorised as Kissy Face blurred the lines between exercise and seduction. The subjects, all female, deployed an overt facial expression, protruding the lips while wearing a limited amount of clothing, to present a highly sexualised self. While these images only accounted for 1 per cent of top posts, they project a sexually idealised female face to signify sex appeal and vitality through the subject's seductive gaze into the camera.

The Glamour Shot is an extension of the Kissy Face trope that focuses more broadly on the subject as a healthy individual. The Glamour Shot accounted for 13 per cent of posts in our sample. In contrast to the Kissy Face trope, the body was the focal point of these images (although, once again, the subject was often highly sexualised, gazing seductively at the camera). These posts only featured women. Rather than being shown undertaking physical activity, the women featured in this type of image were highly made-up and posed in tight active wear to demonstrate their physical fitness, good health and beauty. By representing the female body in overtly sexualised ways, these images resembled the 'fitspo' imagery that disproportionately portrays lean women posing in revealing swimsuits and athletic wear online (Lupton, 2018a). In this regard, the Glamour Shot mimicked the style of feminine beauty commonly featured on the cover of women's magazines in Western media (Kang, 1997).

Another prominent type of image that reinforced ideal notions of health was the Before/After shot. Accounting for 13 per cent of posts, images of this kind showcased the subject's physical transformation through a collage of photos taken before and after a lifestyle change. These posts typically documented women's weight loss over time, signifying the female subjects' physical transformation into 'healthy' subjects as validated by imagery and metrics (e.g., text documenting weight loss). There were also commonalities here between clean eating top posts and the images that regularly feature on fitspo websites. Both privilege a specific body: slim, toned, young and generally white, although often with tanned skin to give the appearance of health and to accentuate muscle tone (Boepple and Thompson, 2016; Lupton, 2018a). In this context, a slender, toned body functions as a synonym for physical fitness and health. The after images typically depicted subjects as visibly healthier, happier and more attractive as a result of their lifestyle change. The Before/After shot here communicates the importance of clean eating practices as techniques to achieve wellness and self-improvement. Collectively, images of this kind reinforce the symbolic logic of adhering to a 'clean' diet and lifestyle.

Individualised conceptions of health

Another significant dimension that characterised clean eating posts collected in our sample was the emphasis on the individual subject rather than groups. Sixty-seven per cent of posts featured an individual subject, in comparison to a mere 9 per cent including two or more people (the remaining 24 per cent featured food). Muscle Presentation, in which subjects were shown flexing their muscles to display physical power and strength, accounted for 42 per cent of clean eating top posts. People in these posts were commonly staged in a gym environment, engaging in exercise to communicate the discipline and effort involved in adhering to a clean eating lifestyle. Here muscle presentation was used to connote more than strength: it was a sign of meritocratic achievement with the self-made 'man' (men overwhelmingly represented in this category) replaced by the self-cured individual – one who has taken their health into their own hands (Baker & Walsh, 2018).

In this context, the body functions as a powerful symbol, signifying the extent to which the subject has self-discipline and self-control. This emphasis on self-discipline and self-control aligns with Western values of responsibilisation and the management of the self (Lupton, 1996, p. 16). The ethos of personal responsibility, and the emphasis on the individual, to which members of the clean eating community subscribe, is also manifest in the 'fitspo' culture online (Lupton, 2018a). However, whereas discourses around weight loss and diet have historically focused on regulating women's bodies, clean eating top posts were represented to a greater degree by men portrayed consuming low calorie, vegetarian food.

While clean eating and meatless diets have generally been viewed as feminine practices, these diets were depicted as a rational choice for men when framed in terms of the individual's health and protection of the environment (Mycek, 2018). The inclusion of hashtags such as #organic, #healthy and #crueltyfree in the captions of these posts upheld the traditional association between reason and masculinity. Other frequently used hashtags that accompanied clean eating top posts reinforced themes of responsibilisation as articulated by the words

#determination, #motivation and #healthy choices. In promoting the idea that health is an individual choice and responsibility, these posts reinforce ideas of 'healthism' that positions health as the result solely of individual behaviour (Crawford, 1980). By focusing on the disciplined individual, the underlying message is that the health solution resides with the individual's determination to 'resist culture, advertising, institutional and environmental constraints, disease agents, or, simply, lazy or poor personal habits' (Crawford, 1980, p. 378). The corresponding hashtags that accompanied clean eating images such as #workhard, #no excuses, #nevergiveup and #healthy choices further obscured the fact that food choices are situated within broader sociocultural, economic and political contexts.

Idealisation

The analytical categories associated with clean eating all point to pertinent examples of idealisation, which highlight instances where the individual will offer impressions of themselves that remain unrealistically ideal. Although presented in highly individualised ways (posts overwhelmingly featuring the individual subject rather than a collective), idealisation works by drawing on and symbolically reinforcing the shared beliefs of the group through visual ritual display. As Goffman (1959, p. 35) contends, such performances presented by an individual 'will tend to incorporate and exemplify the official accredited values of the society'. The Kissy Face, Glamour Shot and Muscle Shot are categories of visual display that offer versions of subjects that are sexually primed and presented for the camera.

These images are idealised because they appear staged for dramatic effect. Although they are generally presented as capturing impulsive and unrehearsed mundane aspects of everyday life, most of the images appear to involve a high degree of staging to communicate heightened signs of health and fitness bound to normative ideals of sexual attractiveness. While the static Before/After shot offers an impression of the body that celebrates the starkest transitions in its physical form, these images are successful idealising media because they structurally juxtapose the before shot with the after image, enabling for explicit comparison that exaggerates the differences displayed. Idealisation points to the ways in which strategies for self-presentation become mediated and how the materiality of digital and face-to-face settings represent different varieties of architecture that stage and shape human behaviour.

In this context, drawing on Goffman's theoretical framework grounds our analysis in Durkheimian notions of social structure and helps to explain how ritualised behaviours and presentational practices associated with clean eating maintain the normative order of society (Collins, 1986; Durkheim, [1912] 1995). The role of rituals in establishing and maintaining the beliefs and practices of the clean eating community is evidenced by homogeneous imagery pertaining to 'clean' foods and its effects on the body. Our analysis demonstrates how even those 'polluting' foods that appear to violate clean eating principles are accepted when substituted with ingredients and practices that adhere to the group's ethical principles and beliefs (e.g., 'cruelty free' desserts made with 'healthy' substitutes). Negative rituals are also deployed to reinforce preferences for clean foods through avoidance practices and via omitting representations of foods that eschew clean eating ideals.

Affirmation ritual

Rituals play an integral role in reinforcing group beliefs (Douglas, [1966] 2002; Durkheim, [1912] 1995). Rituals assume heightened significance on social media given that many of these platforms are designed to encourage users to seek status, attention and affirmation. Goffman's analysis of how acknowledgement and affirmation have become embedded in interpersonal conduct is useful for understanding the emergence and maintenance of digital food cultures on social media. Seemingly inconsequential ritualised interactions have been adapted to digital contexts. In the case of Instagram, the act of liking and commenting on posts can be conceived as affirmation rituals staged in a digital domain. For example, clean eating top posts on Instagram tend to represent specific types of foods, namely, raw produce in its unprocessed form and those so-called 'superfoods' – açaí berries, chia seeds and coconut oil – associated with the clean eating movement. It is these images that are 'liked' and elevated by other users to achieve top post status. These 'superfoods' are considerably more expensive than other varieties. Their validation as healthy and superior to other foods frame clean eating as a largely middle-class pursuit (McCartney, 2016). The same can be said of the types of bodily images validated on Instagram that pay homage to this lifestyle. In all cases, a very specific diet and body type is 'liked' and modelled by others on the platform.

While such acts may appear seemingly unremarkable as a result of their ubiquity, their significance is that they demonstrate what a particular community values and how it organises and elevates particular dimensions of this identity. They have a considerable role in social organisation, as Goffman reminds us about seemingly trivial interactions. Affirmation rituals, such as liking, are intensified in the context of Instagram through the mobilisation and visibility of metrics. The rituals associated with the clean eating movement idealise and project understandings of gendered practice and health as an individual responsibility within social encounters online. Here the degree of affirmation is signified by the number of individuals who have engaged with a given post. In the case of celebrities or influencers with large followings, the degree of affirmation can reach the millions. Indeed, it is the extent to which this interaction becomes asymmetrical that status is elevated and achieved.

Conclusion

The proliferation of digital media in social life has had a profound impact on how food meanings and practices are represented and shared online.

Digital technologies afford users with new modes of self-presentation and the capacity to communicate with online social networks across vast temporal and spatial contexts. Hashtags provide spaces for like-minded individuals to interact and can be used by individuals collectively to enact an alternative food politics. The images shared in these online spaces are strong carriers of meaning. They convey not only discourses around food, but also the self and society more generally.

In this chapter, we have examined how clean eating communities present themselves online by conducting a visual content analysis of top posts pertaining to popular clean eating hashtags on Instagram. Drawing on Durkheim, Douglas and Goffman as a theoretical foundation for this study, we have situated the clean eating movement in cultural ideas of purity and defilement. Concepts of clean and unclean food constitute identity at an individual and collective level by defining accepted eating practices and attempting to regulate behavioural norms. We have suggested that the concept of ritual, foundational to these theoretical approaches, is useful for understanding how status and identity embedded in visual displays encourage rituals of liking, commenting and sharing. This is evidenced by the high degree of homogeneity among clean eating top posts. Even those top posts that appeared to deviate from clean eating principles were shown to align with the community's beliefs. From this perspective, the clean eating movement can be understood in terms of symbolic boundary-maintenance with clean eating structured around beliefs about health and wellbeing.

Clean eating practices are displayed on Instagram to represent an ideal self to a person's social network. Idealisation was apparent in posts featuring the body and food. Although food images were significantly underrepresented in clean eating top posts, the foods that featured in top posts were used by clean eating practitioners to stand in for an idealised self. At the same time, indulgent food that appeared to undermine clean eating principles was idealised when framed in relation to healthy and ethical principles (e.g., organic and cruelty-free produce, or sustainable framing practices). Idealisation was particularly manifest in gender display, with images for the most part reinforcing hegemonic versions of sexualised femininity rather than challenging gender stereotypes. Conversely, conventional associations between masculinity and meat eating were challenged by imagery of muscular men consuming and advocating vegan and vegetarian practices.

While these representations appeared to resist and reframe conventional gender norms, the hashtags that accompanied these images depicted these dietary practices as a rational choice for the individual's health and the environment, thereby upholding the traditional symbolic connection between reason and masculinity. This emphasis on the individual in clean eating top posts aligns with Western conceptions of health as an individual pursuit requiring discipline, sacrifice and commitment on behalf of the subject. Akin to religion, clean eating assigns a moral value to food. The representation of clean eating practices on Instagram reinforces traditional hierarchies by affirming and reinforcing contemporary Western conventions of diet, gender and health. The moral value assigned to food transfers onto the consumer with the body operating as a signifier for idealised conceptions of health, character and status. Those that adhered to a clean eating lifestyle were depicted as superior physically and morally, so that you are what you eat literally and symbolically.

Further reading

Baker, S.A., & Walsh, M.J. (2018). How men are embracing 'clean eating' posts on Instagram. *The Conversation*, 19 June. Retrieved from https://theconversation.com/how-men-are-embracing-clean-eating-posts-on-instagram-97923.

References

Abidin, C. (2014). #In$tagLam: Instagram as a repository of taste, a brimming marketplace, a war of eyeballs. In M. Berry & M. Schleser (Eds.), *Mobile Media Making in the Age of Smartphones* (pp. 119–128). London: Palgrave Pivot.

Alexander, J.C. (Ed.). (1990). *Durkheimian Sociology: Cultural Studies*. Cambridge: Cambridge University Press.

Alexander, J.C. (2004). Cultural pragmatics: Social performance between ritual and strategy. *Sociological Theory, 22*(4), 527–573.

Allen, M., Dickinson, K., & Prichard, I. (2018). The dirt on clean eating: A cross-sectional analysis of dietary intake, restrained eating and opinions about clean eating among women. *Nutrients, 10*(9), 1–11.

Bailey, E.J. (2014). Food Rituals. In P.B. Thompson & D.M. Kaplan (Eds.), *Encyclopedia of Food and Agricultural Ethics* (pp. 945–951). Dordrecht: Springer.

Baker, S.A. (2014). *Social Tragedy: The Power of Myth, Ritual and Emotion in the New Media Ecology*. New York: Palgrave.

Baker, S.A., & Rojek, C. (2019a). *Lifestyle Gurus: Constructing Authority and Influence Online*. Cambridge: Polity.

Baker, S.A., & Rojek, C. (2019b). The Belle Gibson scandal: The rise of lifestyle gurus as micro-celebrities in low trust societies. *Journal of Sociology*. Online first.

Baker, S.A., & Walsh, M.J. (2018). 'Good Morning Fitfam': Top posts, hashtags and gender display on Instagram. *New Media & Society, 20*(12), 4553–4570.

Boepple, L., & Thompson, J.K. (2016). A content analytic comparison of fitspiration and thinspiration websites. *International Journal of Eating Disorders, 49*(1), 98–101.

boyd, d. (2011). Social network sites as networked publics: Affordances, dynamics, and implications. In Z. Papacharissi (Ed.), *Networked Self: Identity, Community, and Culture on Social Network Sites* (pp. 39–58). New York: Routledge.

Brownlie, J., & Shaw, F. (2019). Empathy rituals: Small conversations about emotional distress on Twitter. *Sociology, 53*(1), 104–122.

Burns, T. (1992). *Erving Goffman*. London: Routledge.

Cockburn, A. (1977). Gastro-Porn. *The New York Review of Books*. Retrieved from www.nybooks.com/articles/1977/12/08/gastro-porn/.

Collins, R. (1986). The passing of intellectual generations: Reflections on the death of Erving Goffman. *Sociological Theory, 4*(1), 106–113.

Collins, R. (2004). *Interaction Ritual Chains*. Princeton: Princeton University Press.

Contois, E.J. (2015). Guilt-free and sinfully delicious: A contemporary theology of weight loss dieting. *Fat Studies, 4*(2), 112–126.

Crawford, R. (1980). Healthism and the medicalization of everyday life. *International Journal of Health Services, 10*(3), 365–388.

Döring, N., Reif, A., & Poeschl, S. (2016). How gender-stereotypical are selfies? A content analysis and comparison with magazine adverts. *Computers in Human Behavior, 55*, 955–962.

Douglas, M. ([1966] 2002). *Purity and Danger: An Analysis of Concepts of Pollution and Taboo.* London: Routledge.

Durkheim, É. ([1912] 1995), *The Elementary Forms of Religious Life.* New York: Free Press.

Elias, A., Gill, R., & Scharff, C. (2017). Aesthetic labour: Beauty politics in neoliberalism. In *Aesthetic Labour: Rethinking Beauty politics in Neoliberalism* (pp. 3–49). London: Palgrave Macmillan.

Giddens, A. (1992). *The Consequences of Modernity.* Cambridge: Polity Press.

Gill, R., & Elias, A.S. (2014). 'Awaken your incredible': Love your body discourses and postfeminist contradictions. *International Journal of Media & Cultural Politics, 10*(2), 179–188.

Goffman, E. (1959). *The Presentation of Self in Everyday Life.* New York: Anchor Books.

Goffman, E. (1967). *Interaction Ritual: Essays in Face-To-Face Behavior.* Chicago: Aldine Publishing.

Goodman, D., DuPuis, E.M., & Goodman, M.K. (2012). *Alternative Food Networks: Knowledge, Practice, and Politics.* London: Routledge.

Haidt, J. (2012). *The Righteous Mind: Why Good People Are Divided by Politics and Religion.* New York: Pantheon.

Highfield, T., & Leaver, T. (2016). Instagrammatics and digital methods: Studying visual social media, from selfies and GIFs to memes and emoji. *Communication Research and Practice, 2*(1), 47–62.

Hogan, B. (2010). The presentation of self in the age of social media: Distinguishing performances and exhibitions online. *Bulletin of Science Technology & Society, 30*(6), 377–386.

Instagram (2017). *What are Top Posts on hashtag or place pages?* Retrieved from www.facebook.com/help/instagram/701338620009494?helpref=hc_fnav.

Kang, M. (1997). The portrayal of women's images in magazine advertisements: Goffman's gender analysis revisited. *Sex Roles, 37*(11–12), 979–996.

Kaye, L.K., Malone, S.A., & Wall, H.J. (2017). Emojis: Insights, affordances, and possibilities for psychological science. *Trends in Cognitive Sciences, 21*(2), 66–68.

Lelwica, M.M. (2002). *Starving for Salvation: The Spiritual Dimensions of Eating Problems among American Girls and Women.* Oxford: Oxford University Press.

Lupton, D. (1996). *Food, the Body and the Self.* London: Sage.

Lupton, D. (2018a). Vitalities and visceralities: Alternative body/food politics in new digital media. In M. Phillipov & K. Kirkwood (Eds.), *Alternative Food Politics: From the Margins to the Mainstream* (pp. 151–168). London: Routledge.

Lupton, D. (2018b). *Fat.* London: Routledge.

Marwick, A.E. (2013). *Status Update: Celebrity, Publicity, and Branding in the Social Media Age.* New Haven: Yale University Press.

Marwick, A.E. (2015). Instafame: Luxury selfies in the attention economy. *Public Culture, 27*(1), 137–160.

McCartney, M. (2016). Clean eating and the cult of healthism. *British Medical Journal, 354*, i4095.

Mudry, J. (2018). Nutrition, health, and food: 'What should I eat?'. In K. LeBesco & P. Naccarato (Eds.), *The Bloomsbury Handbook Food and Popular Culture* (pp. 274–285). London: Bloomsbury.

Mycek, M.K. (2018). Meatless meals and masculinity: How veg* men explain their plant-based diets. *Food and Foodways, 26*(3), 1–23.

Nevin, S., & Vartanian, L. (2017). The stigma of clean dieting and orthorexia nervosa. *Journal of Eating Disorders, 5*(1), 37.

Page, R. (2012). The linguistics of self-branding and micro-celebrity in Twitter: the Role of hashtags. *Discourse & Communication, 6*(2), 181–201.

Pate, C.E. (2006). Acknowledgement rituals: The greeting phenomenon between strangers. In J. O'Brien (Ed.), *The Production of Reality: Essays and Readings on Social Interaction* (pp. 169–184). Newbury Park: Pine Forge Press.

Smith, A., & Anderson, M. (2018). Social media use in 2018. *Pew Internet & American Life Project*, 17 September. Retrieved from www.pewinternet.org/2018/03/01/social-media-use-in-2018/.

Tiidenberg, K. (2018). *Selfies: Why We Love (and Hate) Them*. London: Emerald Publishing.

Tiidenberg, K., & Gómez Cruz, E. (2015). Selfies, image and the re-making of the body. *Body & Society, 21*(4), 77–102.

Walsh, M.J. & Baker, S.A. (2017). The selfie and the transformation of the public–private distinction. *Information, Communication & Society, 20*(8), 1185–1203.

Walsh, M.J., & Baker, S.A. (forthcoming). *Clean Eating and Instagram: Purity, Defilement and the Idealisation of Food*.

Walsh, M.J., & Clark, S.J. (2019). Co-present conversation as "socialized trance": Talk, involvement obligations, and smart-phone disruption. *Symbolic Interaction, 42*(1), 6–26.

5

HEALTHISM AND VEGANISM

Discursive constructions of food and health in an online vegan community

Ellen Scott

Introduction

This chapter explores some of the ways vegans discursively construct their diet in relation to health. Online forums represent a crucial site of discourse and meaning construction for vegans, particularly around food. Vegans are a widely dispersed minority who can face hostility in mainstream spaces, and therefore, many find solidarity in digital spaces (Wrenn, 2017a). While veganism extends beyond diet, everyday food choices are the primary way veganism is enacted. As a result, food, nutrition and health are frequent topics of online discussion for vegans. Historically, vegan diets have been widely understood as a radical form of ascetism and connected with weakness and ill-health, reflecting entrenched cultural understandings of meat as vital nourishment (Cole & Morgan, 2011). As such, legitimising vegan diets as healthy and achievable is a significant activity for vegans, both online and in the 'real' world.

While long-standing ideas of veganism as restrictive and unhealthy continue to proliferate, perceptions of veganism are also in flux as the movement experiences a surge in popularity and cultural salience. The number of vegans in Western countries is thought to have doubled in the past decade, currently estimated at between 1 and 5 per cent of the population (Asher, Green, Gutbrod, Jewell, Hale, & Bastian, 2014; Roy Morgan Research, 2016; The Vegan Society, 2016). 'Vegan' has also become a key trend in mainstream food industries (Baum & Whiteman Consultants, 2018; Global Data, 2017; Mintel Group, 2017). While not converting to veganism, a significant number of people are buying vegan cookbooks, choosing vegan meals at eateries, participating in campaigns such as 'Meat Free Monday' and buying plant-based alternatives in a trend dubbed 'flexitarianism' (Budgar, 2011; O'Donnel, 2014; Swanson, 2015). Increased engagement with the culinary aspects of veganism reflects an emergent discourse of veganism as 'empowering, health supportive, and even sexy' (Wright, 2015, p. 42).

Previous research

Sociological inquiries into veganism are primarily concerned with the process of becoming a vegan and motivations for doing so (e.g., Beardsworth & Keil, 1991; Hirschler, 2011; Larsson, Rönnlund, Johansson, & Dahlgren, 2003; McDonald, 2000). Within this literature, health is consistently cited as a key motivator for adopting a vegan diet, along with ethical and environmental concerns. While health is acknowledged as a significant motivator, little is known about how vegans meaningfully understand health or make sense of the perceived health benefits of their diet. This cannot be assumed, as health is a notoriously vague, complex and subjective concept that encompasses a wide variety of meanings which can change over time and place (Cheek, 2006). Noting the broader moralisation of health in society, Beardsworth and Keil speculate that health benefits could represent 'symbolically significant rewards for moral rectitude' for vegans and vegetarians (1991, p. 42).

Christopher, Bartkowski and Haverda (2018) offer further insight in their analysis of how veganism is portrayed in two popular vegan advocacy films. Intended to convert people to veganism, these films are considered a critical medium through which veganism and vegan identity is culturally constructed by 'elite advocates' (Christopher et al., 2018). Two distinct approaches to veganism are observed: 'health veganism' and 'holistic veganism' (Christopher et al., 2018, p. 64). Holistic veganism includes concern for ethics, health and environmental principles, while health veganism is focused solely on health. Health veganism is argued to reproduce hegemonic 'cultural discourses of health, individualism, and responsibility' in the films (Christopher et al., 2018, p. 74).

These discourses can be further understood in terms of healthism. Healthism is a pervasive ideology that equates health with purity, civility and morality (Crawford, 1994). Health is extended beyond the absence of disease to encompass a transcendent, idealised state of physical, mental and emotional perfection under healthism (Crawford, 1980). A key feature of healthism is a singular focus on lifestyle factors – most notably diet – eclipsing the role of environmental, genetic, social and structural influences on health (Shea & Beausoleil, 2012). Responsibility for health is displaced from the state, becoming the sole responsibility of individuals (Crawford, 1994). Industrious repetition of 'healthy' behaviours in all aspects of daily life becomes the expected duty of responsible citizens. Those who cannot or will not achieve a certain standard of health are constructed as irresponsible, indulgent, lazy, uncivilised and morally lacking (Lupton, 2005).

Healthist discourses accord transcendental significance to health, which has become a secular form of salvation associated with purity and morality (Cheek, 2006; Hamilton, Waddington, Gregory, & Walker, 1995). These religious themes are theorised to stem from the Protestant work ethic (Crawford, 1994). Both reinforced by and reinforcing the prevailing political rationality of neoliberalism, healthism is a dominant cultural framework for understanding health and diet in contemporary Western cultures (Crawford, 2006). As such, healthism shapes the 'socially and culturally available array of symbols and meaning' that vegans draw from (Williams, 2004, p. 96). Sociological investigations of healthism are primarily focused on the mainstream health discourses in areas such as government policy, health promotion or popular media, with little attention to how these discourses are negotiated beyond the mainstream.

Greenebaum (2012) finds that vegans focus on health benefits when discussing veganism with outsiders as a 'face saving' strategy. This allows them to side-step the more contentious aspects of veganism, reorienting it towards widely embraced cultural values around healthy eating. Managing social interactions is noted as a significant issue for vegans, who report hostility and microaggressions from non-vegans when eating or socialising together (LeRette, 2014; Twine, 2014). Some attribute this to envy, as vegans are perceived as significantly more virtuous than those who consume animals (MacInnis & Hodson, 2017; Ruby & Heine, 2011). Hostility is also theorised to stem from the symbolic threat veganism poses to dominant cultural norms (MacInnis & Hodson, 2017; Packwood-Freeman & Leventi-Perez, 2012; Twine, 2014). Consumption of animals is deeply embedded in Western cultures as natural and essential (Flail, 2006; Sahlins, 1993). The origins of modern civilisation itself are mythically tied with meat, which acts as a potent symbol of human power and control over nature (Fiddes, 2004; Swinbank, 2002). Vegan rejection of this mythic and deeply embedded cultural symbol is argued to provoke defensive hostility from those invested in normative understandings of animal consumption (Fiddes, 2004; Packwood-Freeman & Leventi-Perez, 2012; Twine, 2014).

In response to mainstream hostility, many vegans seek refuge online. The significance of online spaces for the vegan community is increasingly recognised (Wrenn, 2017a). Vegans are a widely dispersed minority who require supportive social networks to maintain their veganism (Cherry, 2006; Larsson et al., 2003). The internet closes geographic distance, enabling a thriving global vegan community. Blogs, forums and social networking sites have been identified as important locations of vegan community-building, organising, information exchange, cultural transmission, identity-building and meaning construction (Nolasco, 2016; Veron, 2016; Wrenn, 2017a). Additionally, vegans use the internet to promote veganism to a wider audience (Bosworth, 2012; Nolasco, 2016).

Studies of vegans online also expose a darker side to the movement. The operation of white supremacist (Harper, 2011) and fatphobic (Wrenn, 2017a) discourses in online vegan spaces have been revealed. Veganism is generally understood as a left-leaning ideology based on principles of social justice (Wrenn, 2017b). However, the movement is also increasingly perceived as a predominately white, elitist and privileged movement that alienates marginalised groups (Greenebaum, 2018; Harper, 2010, 2013; Polish, 2016; Wrenn, 2017a). Through analysis of online discourses, Harper (2011) and Wrenn (2017a) offer a glimpse into how these issues play out in the everyday lives of vegans online.

Methodology

In this chapter, I adopt a cultural sociological perspective which acknowledges food and health as deeply symbolic domains, embedded with a wealth of complex cultural meaning (Douglas, 1966; Fitzgerald, 1998; Lupton, 1996). Culture is understood as an intricate text underpinned by a semiotic structure of signs and symbols, that is both a pre-structured system of symbols and a reflexive practice for meaning construction (Alexander, 2003; West, 2009). Culture can be studied as a 'social text replete with codes, narratives, genres, and metaphors' (Reed & Alexander, 2009, p. 382). Language, narrative and binary codes are recognised as integral in the construction of cultural meaning, particularly in the diffusely symbolic realms of food and health (Fairclough, 2003; Malesh, 2009; Seidman, 2013).

In order to explore how food is meaningfully constructed in relation to health within vegan culture, a digital discourse analysis was undertaken. Digital discourse analysis is concerned with how meaning is produced through online texts, and with the ways in which these texts and meanings might reflect or interact with larger social and cultural forces (Recuber, 2017), with roots in Fairclough's critical discourse analysis (1995) as well as the work of Altheide (1996) and Ruiz (2009). Discourse is understood in this context as patterned ways of thinking and speaking about things that provide frameworks of meaning through which the world is understood (Fairclough, 2003; Hall, 2001).

For the purposes of my study, naturally occurring discourse within a vegan online discussion forum was selected for analysis. This kind of digital data can be richly detailed and highly meaningful (Skageby, 2015). Insights are drawn by situating forum discourses within the wider cultural context in which they are embedded. The forum, VegChat, was selected based on popularity and frequency of posts, located by selecting the most popular (as per number of unique visits) 'vegan forum' on Google. VegChat is general in scope, facilitating discussions on any vegan-related topic. All threads posted over a three-month period between 1 May 2015 and 31 July 2015 were initially collected, totalling 135 individual threads. A thread consists of a post and any comments the post receives. An initial manual coding was undertaken to identify all threads that contained 'health talk', defined as any discussion of health-related issues, whether tangential or central. This encompassed topics such as diet, nutrition, medication, exercise, weight, fatness, illness or disease, skin and hair health, digestive issues, mental health and wellbeing, energy, and sleep.

A final corpus of 82 threads containing health talk were collated. Most of these were centrally concerned with health-related issues, with around 20 per cent mentioning them tangentially. Health talk occurred consistently throughout the analysis period, occurring in roughly six threads per week. Threads were overwhelmingly text-based but also included some emojis, videos, memes, comics, infographics, photographs and hyperlinks to other websites. Posts varied in length from as short as 30 words to as long as 2,300, with an average of 200 words per post. Each post attracted an average of eight comments, which varied from 25 to 1,200 words in length with an average of 130 words each. In total, the 82 threads consisted of over 100,000 words, nine images, and three videos. These data were thematically analysed to inductively identify key patterns in the way health and food were connected and discussed. Dominant themes were then subject to an extended process of systematic reviewing and refining, with illustrative extracts collated for each theme.

Approval from my university's human research ethics committee was not required for this project, as pre-existing data were collected from a publicly available source with no sign-in requirement. These are considered public space, so consent to observe is not necessary (Convery & Cox, 2012; Warrell & Jacobson, 2014). However, it is important to remember the boundary between public and private can be blurred online. Forums can 'foster a sense of intimacy, community and an expectation of privacy among members' (Orton-Johnson, 2010, p. 15). It is possible posters see their conversations as 'a quiet chat behind closed doors' occurring within the confines of the vegan community (Dawson, 2014, p. 433). To respect the privacy of those involved, defaulting to conventions of anonymity is recommended (Convery & Cox, 2012; Hewson et al., 2017). As such, VegChat is referred to under a pseudonym and users' screen names, and any identifying information including photographs and videos of posters will not be provided in this chapter.

Redefining veganism: rebirth and redemption

A notable finding from my analysis of the data corpus was the high frequency of health-related discussions. Out of the 135 discussion threads initially collected, 82 discussed health: 60 per cent of all threads within the three-month sample period. While VegChat is open to any subject related to veganism, the topic of health took up a majority of discussion space, far outweighing discussions on the ethical or environmental dimensions of veganism. Many contributions by posters on VegChat centred on making claims for the superior healthfulness of a vegan diet. There was a clear ongoing dialogue with mainstream cultural discourses that construct animal-derived foods, most prominently meat, as a vital source of nourishment (Fiddes, 2004).

This preoccupation could be a reaction to the persistent historical construction of veganism as unhealthy. Posters were aware of stereotypes of vegans as weak and unhealthy, and consistently worked to reframe vegan bodies as superior. This was achieved through anecdotes emphasising how much better they felt on a vegan diet:

> I feel healthier than I ever have and my performance as a runner has improved significantly as a result of eating this way.

> I went vegan (I thought temporarily) a few months ago, but because I feel so wonderful, I just never made it back to my meat-eating ways.

Yesterday I ate fully Vegan for the first time (day 1 as Vegan), and I ate about 1,500 calories of good REAL FOOD that keep me full and nourished ... I was ENERGIZED and HAPPY and I felt more awake in the afternoon than I did when I ate meat and other animal products.

I'm so passionate about getting my health back ... I really do feel THAT much better. Even after going on 3 years of eating this way, once in a while I'll do something and think, 'I used to not be able to do this!' Some time back, I was leaning over, scrubbing our bath tub, and realised that my knees, hips, and wrists didn't hurt at all. I remember overwhelming thankfulness coming over me when I remembered how difficult that used to be for me.

Several friends, my parents, my mother in law, and my husband have all taken up eating vegetarian meals, after seeing the way I look/feel after eating this way.

Through these accounts, vegan converts are bestowed with improved physical abilities and a generalised sense of wellbeing, nourished by wholesome vegan food. The traditional association between meat and strength is contested, with the removal of meat instead associated with increased energy and stamina.

Vegan diets were also frequently medicalised, presented as a treatment or cure for a variety of health problems. These ranged from minor issues to more serious diseases:

We discovered that plant-based eating might be the secret to getting rid of [my daughter's] persistent acne. She then joined me in vegan land and has had some noticeable success with her skin.

I know that a lot of people report improved sleep quality on plant-based vegan diets.

When I went vegan my digestion improved dramatically.

I have osteoarthritis in my feet and ankles and eliminating animal products, particularly dairy, has made the most improvement.

Eating this way helped my heart and kidneys.

I'm Type 2 diabetic, non-medicated at the moment ... and my doctor is thankfully supportive of my efforts to better my health without meds [through a vegan diet and exercise].

In the 1940s Walter Kempner cured diabetes with a diet consisting solely of white rice, fruit, fruit juice, added table sugar, and a multivitamin.

John Mcdougall continues to cure type II diabetes with a starch-based 10% fat vegan diet.

I've seen a plant-based diet bring a man back to life who had one foot in the grave.

Personal experience was the most commonly cited evidence in posts for the healing properties of veganism. Narratives of vegan health transformation were enthusiastically received, described as 'inspirational'. Through these anec-dotes, conversion to veganism was constructed as a transformative experience, evoking quasi-spiritual themes of redemption and renewal. The vegan self is reinvented through food: healed, healthy and happy. This secular salvation is achieved through the body rather than the soul (Hamilton et al., 1995). Building on themes of renewal and rebirth, vegan diets were also presented as having dif-fusely anti-aging properties:

I think I look better at 42 [after 14 years as a vegan] than I did at 20 years old, hands down. I am more toned and stronger, more active, have more energy... I am still mistaken for being in my twenties in some photos.

One poster shared two photographs of herself taken eight years apart, titling her post 'Proof, it's the FOOD! I'm getting younger'. She received the following responses:

Awesome, you look great! And yep food is huge when it comes to slowing the aging process I'm 36 and still get mistaken for being 25, I always tell people it's the diet haha.

You look cute in both photos! You've really transformed your health, though!

You really do look younger and healthier in the later photo. That's just more evidence that a whole-food, plant-based diet is the healthiest diet. Please continue to keep us updated as to how your diet is positively affect-ing your health. You are an inspiration to all of us!

For these posters, a youthful appearance further cements the health benefits of ve-ganism, blurring aesthetic ideals with ideas of health. Straightforward connections between inner health and outer physical appearance were routinely made, particu-larly in relation to body weight. Weight loss, which was consistently associated with veganism, was frequently offered as evidence of improved health, drawing on broader cultural discourses of fat and weight as signifiers of health (Lupton, 2013).

While positive changes to health were attributed to veganism, negative changes were not. Some posters connected excessive weight loss and fatigue with their new vegan diet. When this occurred, responding comments placed responsibility for this on individuals rather than veganism itself. These posters were labelled as 'inexperienced vegans' who had not learned the ropes of vegan diets yet. Long lists of high calorie vegan foods and meal plans emphasised as 'simple' or 'easy' were provided to these posters. This aligns with previous research by Sneijder and Te Molder, which found that vegans employ their identity as 'ordinary' people to refute ideas of veganism as being complicated or unhealthy (2009). There was some acknowledgement that not all vegans or vegan foods were healthy. One poster reflected that a vegan diet can theoretically consist of 'soda and chips or soy burgers with soy cheese' and that such a diet 'might fail to promote healing'. However, accounts were overwhelmingly focused on the positive health benefits of veganism.

Animal-derived pollutants

While vegan foods were firmly associated with good health, animal-derived foods were consistently positioned in binary opposition to vegan food as dangerous to health:

Animal protein damages your body slowly. Many of our diseases today are caused partially or fully by eating Animal protein.

Animal protein diet that destroys my arteries.

Animal protein – including that from eggs – is hard on the kidneys and encourages calcium loss. Instead of building you up, it is gradually tearing you down.

One of the dangers of meat is that hormones do not disintegrate or go away. It moves up the food chain. Given that humans are on top of the food chain. The hormones accumulates in the human body, it does not go away. Hormones disrupt the endocrine system.

Non-vegan foods make me bloated, sluggish, and depressed.

Yesterday I ate fully Vegan for the first time … That was the first time in months I didn't feel like garbage for what I put in my mouth.

I want to get rid of all animal products because those foods make me feel disgusting and sluggish.

I am slowly recovering from having an unhealthy diet before with my new vegan ways.

A Master Cleanse might be in order to cleanse [animal products from the body].

Conversion to veganism was framed as a process of purification, 'detoxing' from non-vegan foods which were constructed as hazardous pollutants, 'insidiously transforming the eater from within' (Fischler, 1988, p. 281). Ingestion of these pollutants was seen to cause both physical and emotional distress for the eater, leaving them lethargic, depressed and vulnerable to disease.

While these contributions were often couched in objective and scientific language, the binary categorisation of vegan as healthy and non-vegan as unhealthy is symbolically charged. Health is a key marker of value in Western societies and strongly tied with notions of civility (Seidman, 2013). Foods designated as 'good' for health are considered the most moral and civilised to eat, while those considered 'bad' for health are immoral and uncivilised (Lupton, 2005). This binary code extends beyond foods to encompass the people who consume them, analogously imparting the properties of food onto the eater (Fischler, 1988; Sahlins, 1993). Through a positive association with health vegans are constructed as civilised, while non-vegans are cast as the uncivilised other, reinforcing moral boundaries around the movement.

Individual responsibility

The redemptive narrative of veganism was strongly underscored by healthist, neoliberal discourses which position health within a highly individualised framework of risk and responsibility. Good health was routinely presented in VegChat as something posters had control over through education and, ultimately, behaviour. This is clearly illustrated in the way that food choices are strongly connected to health outcomes (as documented above). Bodies were treated as highly malleable. One poster shared a photograph of a '68-year-old long term vegan'. She is posed in a bikini with smooth taut skin and well-defined muscles, giving the impression of someone several decades younger. Her appearance is attributed to her lifestyle and the poster confidently states that anyone can look like her at 68 if they 'eat right, exercise, and don't get too much sun'.

Health advice in the forum focused heavily on lifestyle:

> Walnuts and fresh produce could be helpful in reducing inflammation.

> Sounds like you need to eat more whole plant foods. Dark green veggies, fruits, other veggies, nuts, seeds, whole grains, legumes, delicious sweet potatoes. Drink lots of water, get a good nights sleep, get some exercise. Take a b12 supplement, get some sun or take a vitamin d supplement.

> Removing soda pop from your diet and replacing it with smoothies or plant milks or water will go a long way in helping with weight loss and feeling better.

Health was framed through these discussions as a routine accomplishment achieved through the daily repetition of healthy behaviours, placing the responsibility for health solely on individuals. This was almost unanimously presented as empowering. Forum members revelled in perceptions of unprecedented control over their health. While diet and exercise were frequent topics of discussion around health, the environmental, genetic and structural factors that can impact health were rarely acknowledged. In the sample, genetic factors for disease were acknowledged only three times. Environmental and structural impacts on health were even less recognised, with only one comment in the sample acknowledging them. Risks to health were predominantly understood in individualised behavioural terms.

Discourses of individual responsibility carried an in-built sense of justice in the forum. One poster, for example, said they hoped omnivores 'enjoy the diabetes, heart disease and cancer that the animals give you in revenge for eating them'. The capacity for victim-blaming becomes apparent here. When health is framed as the product of individual choices, those who are unable to achieve good health are inherently at fault. This carries moral implications in a society that values the pursuit of health above all else. People who are chronically ill or disabled are marginalised by these discourses as morally lacking, irresponsible citizens.

Conclusion

My analysis of VegChat threads found that the forum provided space for vegans to collaboratively contest and rework traditional understandings of food and health. Through shared narratives of vegan conversion, veganism was constructed as empowering, joyful and health supportive. This always occurred in dialogue with wider cultural discourses, which shape how vegans understand the world and define the repertoire of discourses available to them. This group storytelling can be understood as a form of movement activity (Malesh, 2009). More than just stories or personal accounts, narratives are a key way we assign meaning and make sense of the world (Smith, 2010). By collaboratively weaving these narratives throughout the forum, posters are 'defining and fortifying [their] collective identity' (Malesh, 2009, p. 133). Through an appeal to the sacred status of health these narratives construct an idealised vegan identity of renewal and redemption.

A binary division opposing vegan food and animal-derived food operated in VegChat. This binary was symbolically significant, drawing on healthist ideas to construct vegans as civilised, pure and morally superior. Discourses of food and health in VegChat were significantly healthist in nature. An idealised, all-encompassing conception of health predominated the forum (Crawford, 1980). As well as having high expectations for health, posters demonstrated high levels of health awareness, spending considerable time and effort seeking out and sharing health information, and were engaged in a continual process of reflection around health behaviours. Health was also imbued with quasi-spiritual properties. Responsibility for health was highly individualised, with environmental, genetic or structural factors impacting health largely unacknowledged. Bodies were constructed as highly malleable objects that could be worked upon through food.

In the VegChat posts, food was also treated as medicine, capable of healing a range of maladies and diseases. Food has become profoundly medicalised in contemporary culture, closely associated with health, illness and disease (Lupton, 2005). This medicalisation of food is part of a broader extension of medical jurisdiction into all aspects of daily life in healthist, neoliberal cultures (Conrad, 1992; Crawford, 1980). By constructing a straightforward relationship between food and health outcomes, those who have less control over health outcomes due to race, class, disability or chronic illness are marginalised. This is problematic for the movement as it further contributes to the elitist and privileged image of veganism.

Further reading

Asher, K., & Cherry, E. (2015). Home is where the food is: Barriers to vegetarianism and veganism in the domestic sphere. *Journal for Critical Animal Studies, 13*(1), 66–91.

Conn, S.G. (2015) Guess who's coming to dinner? Frames, identities, and privilege in the U.S. vegetarian and vegan movement. College of Social Sciences and Public Policy, Florida State University, USA.

LeBesco, K. (2011). Neoliberalism, public health, and the moral perils of fatness. *Critical Public Health, 21*(2), 153–164.

References

Alexander, J.C. (2003). *The Meanings of Social Life.* New York: Oxford University Press.

Altheide, D. (1996). *Qualitative Media Analysis.* Thousand Oaks, CA: Sage.

Asher, K., Green, C., Gutbrod, H., Jewell, M., Hale, G., & Bastian, B. (2014). *Study of current and former vegetarians and vegans, Humane Research Council.* Retrieved from https://faunalytics.org/wp-content/uploads/2015/06/Faunalytics_Current-Former-Vegetarians_Full-Report.pdf.

Baum & Whiteman Consultants. (2018). *Consultants predict 11 hottest food & beverage trends in restaurant & hotel dining for 2018.* Retrieved from https://docs.wixstatic.com/ugd/0c5d00_90935d6fda344991a8fc2452eb112c83.pdf.

Beardsworth, A., & Keil, T. (1991). Health-related beliefs and dietary practices among vegetarians and vegans: A qualitative study. *Health Education Journal, 50*(1), 37–44.

Bosworth, B. (2012). *Spreading the Word: Communicating about Veganism.* Boulder, CO: University of Colorado Boulder.

Budgar, L. (2011). Veganism on the rise among health-conscious consumers, *National Foods Merchandiser, 32*(6).

Cheek, J. (2006). Healthism: A new conservatism? *Qualitative Health Research, 18*(7), 974–982.

Cherry, E. (2006). Veganism as a cultural movement: A relational approach. *Social Movement Studies, 5*(2), 148–161.

Christopher, A., Bartkowski, J.P., & Haverda, T. (2018). Portraits of veganism: A comparative discourse analysis of a second-order subculture. *Societies, 8*, 55–76.

Cole, M., & Morgan, K. (2011). Vegaphobia: Derogatory discourses of veganism & the reproduction of speciesism in UK national newspapers. *The British Journal of Sociology, 62*(1), 134–153.

Conrad, P. (1992). Medicalization and social control. *Annual Review of Sociology*, *18*, 209–232.

Convery, I., & Cox, D. (2012).A review of research ethics in internet-based research. *Practitioner Research in Higher Education*, *6*(1), 50–57.

Crawford, R. (1980). Healthism and the medicalization of everyday life. *International Journal of Health Services*, *10*(3), 365–389.

Crawford, R. (1994). The boundaries of the self and the unhealthy other: Reflections on health, culture and AIDS. *Social Science and Medicine*, *38*(10), 1347–1365.

Crawford, R. (2006). Health as a meaningful social practice. *Health*, *10*(4), 401–420.

Dawson, P. (2014). Our anonymous online research participants are not always anonymous. Is this a problem? *British Journal of Educational Technology*, *45*(3), 428–437.

Douglas, M. (1966). *Purity and Danger*. London: Routledge.

Fairclough, N. (1995). *Critical Discourse Analysis: The Critical Study of Language*. London: Longman.

Fairclough, N. (2003). *Analysing Discourse*. New York: Routledge.

Fiddes, N. (2005). Meat: A Natural Symbol. London: Routledge.

Fischler, C. (1988). Food, self, and identity. *Social Science Information*, *27*(2), 275–292.

Fitzgerald, F.T. (1998). The tyranny of health. *Sounding Board*, *331*(3), 196–198.

Flail, G.J. (2006). *The sexual politics of meat substitutes* (dissertation). College of the Arts and Sciences, Georgia State University, USA.

Global Data. (2017). Top trends in prepared foods 2017: Exploring trends in meat, fish and seafood; pasta, noodles and rice; prepared meals; savory deli foods; soup; and meat substitutes. Retrieved from www.reportbuyer.com/product/4959853/top-trends-in-prepared-foods-2017-exploring-trends-in-meat-fish-and-seafood-pasta-noodles-and-rice-prepared-meals-savory-deli-food-soup-and-meat-substitutes.html.

Greenebaum, J. (2012). Managing impressions: 'Face-saving' strategies of vegetarians and vegans. *Humanity & Society*, *36*(4), 309–325.

Greenebaum, J. (2018). Vegans of color: Managing visible and invisible stigmas. *Food, Culture & Society*, *21*(5), 680–697.

Hall, S. (2001). Foucault: Power, knowledge and discourse. In M. Wetherell, S. Taylor & S.J. Yates (Eds.), *Discourse Theory and Practice* (pp. 72–81). London: Sage.

Hamilton, M., Waddington, P., Gregory, S., & Walker, A. (1995). Eat, drink and be saved: The spiritual significance of alternative diets. *Social Compass*, *42*(4), 497–511.

Harper, A.B. (2010). Race as a 'feeble matter' in veganism: Interrogating whiteness, geopolitical privilege, and consumption philosophy of 'cruelty-free' products. *Journal for Critical Animal Studies*, *8*(3), 5–27.

Harper, A.B. (2011). Veganporn.com & 'sistah': Explorations of whiteness through textual linguistic cyberminstrelsy on the internet. In D. Wachanga (Ed.), *Cultural Identity and New Communication Technologies: Political, Ethnic and Ideological Implications* (pp. 235–255). Hershey, PA: IGI Global.

Harper, A.L. (2013). *Vegan consciousness and the commodity chain: On the neoliberal, afrocentric, and decolonial politics of 'cruelty-free'* (dissertation). University of California.

Hewson, C., Buchanan, T., Brown, I., Coulson, N., Hagger-Johnson, G., Joinson, A., … Oates, J. (2017).*Ethics Guidelines for Internet-mediated Research*. Leicester: British Psychological Society.

Hirschler, C.A. (2011). 'What pushed me over the edge was a deer hunter': Being vegan in North America. *Society & Animals*, *19*, 156–174.

Larsson, C.L., Rönnlund, U., Johansson, G., & Dahlgren, L. (2003). Veganism as status passage: The process of becoming a vegan among youths in Sweden. *Appetite*, *41*, 61–67.

LeRette, D.E. (2014). Stories of microaggressions directed toward vegans and vegetarians in social settings (PhD dissertation). Fielding Graduate University, USA.

Lupton, D. (1996). *Food, the Body and the Self.* London: Sage.

Lupton, D. (2005). Lay discourses and beliefs related to food risks: An Australian perspective. *Sociology of Health & Illness, 27*(4), 448–467.

Lupton, D. (2013). *The Imperative of Health.* London: Sage.

MacInnis, C.C., & Hodson, G. (2017). It aint easy eating greens: Evidence of bias toward vegetarians and vegans from both source and target. *Group Processes & Intergroup Realtions, 20*(6), 721–744.

Malesh, P. (2009). Sharing our recipes: Vegan conversion narratives as social praxis. In S. McKenzie & P.M. Malesh (Eds.), *Active Voices: Composing a Rhetoric of Social Movements* (pp. 131–145). Albany: State University of New York Press.

McDonald, B. (2000). 'Once you know something, you can't not know it': An empirical look at becoming vegan. *Society and Animals, 8*(1), 1–23.

Mintel Group. (2017). *Vegan food launches in Australia grew by 92% between 2014 and 2016.* Retrieved from www.mintel.com/press-centre/food-and-drink/vegan-food-launches-in-australia-grew-by-92-between-2014-and-2016.

Nolasco, A.A. (2016). *Vegan lifestyle on Facebook: An online ethnographic study* (dissertation). University West, School of Business, Economics and IT, Division of Informatics, Sweden.

O'Donnel, K. (2014). Become a flexatarian. *Mother Earth News, 264*, 42–48.

Orton-Johnson, K. (2010). Ethics in online research: Evaluating the ESRC framework for research ethics categorisation of risk. *Sociological Research Online, 15*(4), 13–17.

Packwood-Freeman, C., & Leventi-Perez, O. (2012). Pardon your turkey and eat him too. In J. Frye & M. Bruner (Eds.), *The Rhetoric of Food: Discourse, Materiality, and Power* (pp. 103–120). New York: Routledge.

Polish, J. (2016). Decolonizing veganism: On resisting vegan whiteness and racism. In J. Castricano & R.R. Simonsen (Eds.), *Critical Perspectives on Veganism* (pp. 373–391). Switzerland: Palgrave MacMillan.

Recuber, T. (2017). Digital discourse analysis: Finding meaning in small online spaces. In J. Daniels, K. Gregory & Cottom, T.M. (Eds.), *Digital Sociologies* (pp. 47–60). Bristol: Polity Press.

Reed, I., & Alexander, J.C. (2009). Cultural sociology. In B.S. Turner (Ed.), *The New Blackwell Companion to Social Theory* (pp. 378–390). Chichester: Blackwell.

Roy Morgan Research. (2016). The slow but steady rise of vegetarianism in Australia. Retrieved from www.roymorgan.com/findings/vegetarianisms-slow-but-steady-rise-in-australia-201608151105.

Ruby, M.B., & Heine, S.J. (2011). Meat, morals, and masculinity. *Appetite, 56*, 447–450.

Ruiz, J. (2009). Sociological discourse analysis: Methods and logic. *Forum: Qualitative Social Research, 10*(2), 1–30.

Sahlins, M. (1993). Food as a symbolic code. In J.C. Alexander & S. Seidman (Eds.), *Culture and Society: Contemporary Debates* (pp. 94–101). New York: Cambridge University Press.

Seidman, S. (2013). Defilement and disgust: Theorizing the other. *American Journal of Cultural Sociology, 1*(1), 3–25.

Shea, J.M., & Beausoleil, N. (2012). Breaking down 'healthism': Barriers to health and fitness as identified by immigrant youth in St. John's, NL, Canada. *Sport, Education and Society, 17*(1), 97–112.

Skageby, J. (2015). Interpreting online discussions: Connecting artifacts and experiences in user studies. *The Qualitative Report, 20*(1), 115–129.

Smith, T. (2010). Discourse and narrative. In R. Hall, L. Grindstaff & M.C. Lo (Eds.), *Handbook of Cultural Sociology*. New York: Routledge.

Sneijder, P., & Te Molder, H.F.M. (2009). Normalizing ideological food choice and eating practices. Identity work in online discussions on veganism. *Appetite, 52,* 621–630.

Swanson, C. (2015). The Best-Selling Cookbooks of 2014. *Publishers Weekly.* Retrieved from www.publishersweekly.com/pw/by-topic/industry-news/cooking/article/65184-the-bestselling-cookbooks-of-2014.html.

Swinbank, V.A. (2002). The sexual politics of cooking: A feminist analysis of culinary hierarchy in Western culture. *Journal of Historical Sociology, 15*(4), 463–494.

Twine, R. (2014). Vegan killjoys at the table – contesting happiness and negotiating relationships with food practice. *Societies, 4*(4), 624–639.

Vegan Society, The (2016). *Find out how many vegans are in Great Britain.* Retrieved from www.vegansociety.com/whats-new/news/find-out-how-many-vegans-are-great-britain.

Veron, O. (2016). From seitan bourguignon to tofu blanquette: Popularising Veganism in France with food blogs. In J. Castricano & R. Simonsen (Eds.), *Critical Perspectives on Veganism,* (pp. 287–305). Switzerland: Palgrave Macmillan.

Warrell, J.G., & Jacobsen, M. (2014). Internet research ethics and the policy gap for ethical practice in online research settings. *Canadian Journal of Higher Education, 44*(1), 22–37.

West, B. (2009). Cultural social theory. In A. Elliot (Ed.), *The Routledge Companion to Social Theory* (pp. 188–202). Abingdon: Routledge.

Williams, R.H. (2004). The cultural constraints of collective action: Constraints, opportunities, and the symbolic life of social movements. In D. Snow, A.A. Soule & H. Kriesi (Eds.), *The Blackwell Companion to Social Mov*ements (pp. 91–115). Oxford: Blackwell.

Wrenn, C.L. (2017a). Fat vegan politics: A survey of fat vegan activists' online experiences with social movement sizeism. *Fat Studies, 6*(1), 90–102.

Wrenn, C.L. (2017b). Trump veganism: A political survey of American vegans in the era of identity politics. *Societies, 7*(4), 32.

Wright, L. (2015). *The Vegan Studies Project: Food, Animals, and Gender in the Age of Terror.* Athens, GA: University of George Press.

6

WORKING AT SELF AND WELLNESS

A critical analysis of vegan vlogs

Virginia Braun and Sophie Carruthers

Introduction

Food does more than simply feed us. Food, and what and how we eat, are political, cultural and social practices; they are also structured by attributes such as age, social class, ethnicity, race and gender. The idea and practice of eating a particular restricted diet for personal health and wellbeing, or environmental, or ethical and/or religious/cultural reasons, has a long history. This intersects with contemporary neoliberal Western contexts, where identity is enmeshed in – and indeed done through – consumption, and through display work via social media. In these sociocultural milieu, 'healthism' (Crawford, 1980) – the individual pursuit of health and, now, 'wellness' – is part of the required work for the contemporary neoliberal subject. In the context of a privileged consumer seeking (ultimate) wellness, 'healthy eating' has transmogrified into increasingly particularised styles and practices, meshed in with moral claims (Delaney & McCarthy, 2014). In this chapter, we take vegan vlogs presented on YouTube as a site to explore contemporary social media meaning-making related to food and dietary restriction, self and wellness.

Healthy eating and veganism

In Western countries, food and eating have often fallen within the scope of public health, with 'healthy eating' guidelines developed at the population level (e.g., Ministry of Health, 2013; NHS, 2016). The specifics of such guidelines shift and evolve, but food tends to be broken down into broad food 'groups' (e.g., proteins, fats, fruit and vegetables), with proportional amounts of each group advised. Despite relatively standardised guidelines, practice does not always align – for various reasons (e.g., Finn, 2011) – and contemporary common-sense articulates a 'crisis of obesity', with public discourse dominated by headlines related to obesity, diet and un/healthy eating. This 'crisis' is often attributed to poor dietary practice: lay (alongside scientific) explanations often focus on overconsumption – of high fat foods, of processed foods, of complex carbohydrates and sugars (e.g., *The Guardian*, 2019).

Definitions of what is food, and what value it holds, are powerful contemporary ontological projects. Popular authors such as Michael Pollan (e.g., 2008) have differentiated between less- or un-processed food and highly processed 'edible foodlike substances' (2008, p. 1) – urging a return to (imagined) eating ways of the past. Food has been discursively constructed into 'good' vs 'bad', 'healthy' vs 'unhealthy', meshing morality with how we eat (Delaney & McCarthy, 2014; Henderson, Ward, Coveney, & Taylor, 2009; Sassatelli, 2004). Until recently, (saturated) fats occupied a position of nutritional pariah, with very limited intake recommended for healthy eating, but this advice has become contested (e.g., Groopman, 2017). Indeed, within much contemporary discourse, sugar (especially sucrose which is 'refined') has recently come to occupy a common-sense position as so bad for us that it might even be 'poison' (e.g., Samadi, 2017), but (ontological) contestation remains (e.g., Levinovitz, 2015).

Within this discursive space, a plethora of particularised diets have appeared, promoting very specific or restricted ways to eat, for health and 'wellness', for weight loss – with thinness often a proxy for health (Finn, 2011) – and, sometimes, for planetary good. Some of these have evolved from earlier diets; some offer new takes on 'healthy eating'. These include so-called paleo, ketogenic, sugar-free, clean, raw, raw till 4, 5:2 and 16:8 diets/ways of eating. Some dietary styles developed to address particular medical conditions (e.g., gluten-free for Coeliac disease) have been far more widely taken up: gluten, like (refined) sugar, has been re-ontologised from a neutral and often unknown substance to widely understood as suspect – if not actually inherently bad for us (Levinovitz, 2015).

One thing that unites these 'diets' is elimination or restriction. Unlike many Western government-approved healthy eating guidelines, which promote variety and 'balance', these diets tend to promote the exclusion of whole food groups (e.g., refined sugar, or grains, in the 'paleo' diet), or food processes (e.g., cooking, in a 'raw' diet), or food at particular times (e.g., consuming food only within an eight-hour window in 16:8). We collectively term these specific eating practices 'wellness diets', because they evoke an ideal of optimal health, achievable through the correct diet, and because they often merge into other aspects of living. Both scholars and food writers have expressed concern that such restricted ('healthy') diets can act as socially acceptable masks for what are problematic eating practices (Tandoh, 2016), with pathologies of eating shifting 'obsession' from thinness to health (Dejmanee, 2016; McCartney, 2016). This concern is echoed in the articulation of a new, not yet fully researched (Missbach & Barthels, 2017), 'eating disorder' termed orthorexia nervosa, characterised as an obsession with eating 'healthy', 'natural' and 'pure' foods (see Bratman, 2017).

Veganism – a diet involving the avoidance of animal-based products, including meat, seafood, dairy products and eggs – offers a potentially different mode of dietary consumption and restriction. Although not new (the term was first coined in 1944; The Vegan Society, nd), the contemporary context has seen substantial increases in veganism (Hancox, 2018; Harrington, Collis, & Dedehayir, 2019), as concerns about animal welfare have aligned with concerns about the environmental impacts of food production, particularly the carbon- and water-use intensity of meat and dairy production. Increasingly rebranded as plant-based eating, the discursive construction of veganism has itself shifted dramatically in this contemporary context (Harrington et al., 2019), and veganism occupies a contested space between dietary mode and 'lifestyle' (Bertella, 2018).

The notion of (vegan) consumption as lifestyle aligns with wellness diets, with an often very individualised orientation (Haenfler, Johnson, & Jones, 2012). Meshing with dominant neoliberal modes, where choice and responsibility for health are sited within (decontextualised) individuals (Riley, Evans, & Robson, 2018), food 'choices' become 'personal' ones (Sandal, 2018), evoking an agent making choices within free and accessible food and eating environments. Such interpretative frameworks elide the sociopolitical context of food and eating (Goodman, Johnston, & Cairns, 2017), and indeed the capitalist drivers at work in contemporary eating 'choices' – sometimes captured in the term 'Big Food' (Williams & Nestle, 2018). In contrast, critical scholars have theorised food and eating as part of a 'complex web of power' (Finn, 2011, p. 39), and there are deep and complex intersections of race and class (and gender) in how we eat; the potential privilege inherent in excluding certain food groups or types from one's diet has been noted (Bailey, 2007; Mycek, 2018). Visible veganism tends to be dominated by whiteness, and women, no doubt reflecting connections between masculinity and meat consumption (see Greenebaum, 2017; Greenebaum & Dexter, 2018; Harper, 2013; Harrington et al., 2019; Mycek, 2018; Ruby & Heine, 2011).

Representationally, veganism intersects many domains across our interests in the production of logics of truth, identity and morality around food and 'healthy eating'. Fitting with Wright's (2015) framing of 'vegan studies', we treat 'vegan' and 'healthy' eating as constructed objects, located within an understanding of (all aspects of) food and food representation as political (Goodman et al., 2017). Our theoretical orientation in interrogating vegan vlogs is to explore the ontological shaping of ways of eating, with an interest in the sociopolitical implications of discourse and representation.

Social media, food cultures and healthy eating

Coincidental with, but deeply enmeshed in, neoliberal and postfeminist contexts, has been the rise and expansion of social media. Web 2.0 allowed emergence of the 'participatory web' (Song, 2010), featuring user-generated/interactive content and interactivity that blurs the boundaries between online or 'virtual' and material or 'real' life, and the lines between producer and consumer/user (Lewis, 2018; Lupton, 2015). Central to this 'participatory' culture (see Beer, 2013; Jenkins, 2014) has been a 'sharing ethos,' where the ease of content sharing has produced an interactive and communicative realm between users (Lupton, 2017). In this space, 'ordinary' persons become entrepreneurial subjects (Bandinelli & Arvidsson, 2013), 'micro-celebrities' telling a particular story of or from their life, in a way that is often promoted as authentic, real and unmediated (Berryman & Kavka, 2017; Senft, 2008).

Social media platforms have increased access to varied constructions around health and eating, conveyed through combinations of text, visuals and audio (Lupton, 2017). Such online spaces not only discursively produce the objects of focus, but also allow for the production of (collective) identities (Schneider, Eli, McLennan, Dolan, Lezaum, & Ulijaszek, 2019) and a felt emotional connection (Berryman & Kavka, 2017). This has implications for both public health (e.g., see Lupton, 2015) and individual practice(s), as it disrupts traditional health information authority and expertise, and shifts engagement with the various sources of 'advice' available to us. Online spaces have shifted some authority and understandings of expertise from professionals to the lay public, producing the 'expert patient' who uses this accumulated 'knowledge' to resist mainstream models of health (Fox & Ward, 2008). Such digital cultures have been theorised as deeply meshed into both consumerism and neoliberal logics (Khamis, Ang, & Welling, 2017), reinforcing the message and expanding the scope to be a self-directing (healthist) agent, responsible for one's own wellbeing.

Just as media in general have been 'thoroughly imbricated in how contemporary food politics are imagined, enacted and appropriated' (Phillipov, 2019, p. 1), social media are a vital scholarly focus for those interested in questions of food, politics and practice. Online food cultures are vast and popular; supposedly authentic narratives of food/eating have proliferated across social media (Koch, 2017). The micro-celebrity afforded some within these spaces has been enormous and contributed profoundly to a (re)shaping of discourse related to food and eating (Goodman et al., 2017; Sandal, 2018). Wider media have proclaimed that 'Instagram changed the food we eat' (Lee, 2017); YouTube may have changed the way we cook (Lewis, 2018). British wellness entrepreneur Ella Mills demonstrates the power and potential of online spaces and celebrity to bleed into material domains and affect practice in contemporary contexts (Goodman et al., 2017). Here, the apparent authenticity in an individual's experiential authority and the 'life-changing' and 'health' claims of wellness discourse collide to provide a perfect storm for digital success. Ella Mills' brand 'Deliciously Ella' grew from food blogging and Instagramming to best-selling cookbooks, delis and commercial food production – highlighting the power and potential of 'wellness capitalism' (Khamis et al., 2017; Koch, 2017; Lewis, 2008). However, this is no rags-to-richness neoliberal success story; much pre-existing privilege and social capital cocoons Deliciously Ella.

Vegan vlogs

Blogs (Hookway, 2017; Koch, 2017) and vlogs (video blogs), their audio-visual sibling (Molyneaux, O'Donnell, Gibson, & Singer, 2008), offer particular spaces for food/eating representation and discourse. Vlogs, disseminated online, offer somewhat embodied storytelling, and have been described as occupying an 'in-between' space of simultaneously private and public (Lange, 2007). As a personal narrative, blogs/vlogs appear to offer authorial agency. However, as comments can feed into subsequent content (Kaplan & Haenlein, 2010), they are also social interactive spaces, a potential conversation of sorts between producer and audience. Blogs/vlogs offer a useful space to explore personal (but public) accounts of 'everyday life' in an ever-changing present (Hookway, 2017).

Broadly interested in contemporary healthy eating food movements and digital lives, we chose vegan vlogs as a site to explore constructions around identity, food, health and eating. We occupy various positions in relation to the topic, as eating non-/vegan, as non-/consumers of 'healthy eating' social media, and as sociocultural critics of gender, embodiment and health. After broad perusal of vegan vlogs on YouTube, we focused data collection on two seemingly popular formats: 'What I Eat in a Day/Week' and 'My Vegan Story/Journey'.

Our YouTube search was conducted in one go, and on a 'clean' computer, to avoid any algorithm 'bias' produced by either of our past searches, to mimic the search returns that someone new to vegan vlog searching might encounter. Over six million hits were produced for each search. In total, 30 vlogs were selected (20 'what I eat'; 10 'story'), with sampling determined by popularity and explanatory richness. We selected the first 20 and 10 'hits' with at least 10,000 subscribers and/or likes, *and* where reasons around eating were articulated in the vlog. We used these criteria to garner richer explanatory accounts, and exclude 'what I eat' vlogs that simply listed foods consumed.

The 'What I Eat' vlogs were 'diary' style, and usually featured the producer checking in at mealtimes throughout the day to present what they have eaten, sometimes with 'snack' reports. Clips often include meal preparation or sourcing, followed by visuals of the food and commentary by the producer, plus features like music. Reflecting increasing 'professionalisation' in social media (Phillipov, 2019), they are clearly 'produced,' a format that began a rise to popularity in 2010 (Sandal, 2018); commercial click-links are often provided through the description box. 'My vegan story' vlogs tended to be longer and more meaning-rich, with the vlogger building a story of why they came to eat in that way, often offering commentary on veganism, 'health' and being 'healthy'. In these, the producer usually sat down and spoke directly to camera, as if mimicking a face-to-face conversation with the viewer (Sandal, 2018).

Although digital cultures set up exciting and important possibilities for research, social media research ethics require particular consideration (e.g., Hookway, 2008; Moreno, Goniu, Moreno, & Diekema, 2013). Given the 'outward' facing orientation of vlogs, and YouTube's space as a public sharing forum, but where users can create 'private' videos (Lange, 2007; Moreno et al., 2013), we followed Hook-way's (2008, p. 177) 'fair game – public domain' approach, where content published without restriction is understood to have low expectation of privacy. However, fitting our analytic interest in societal-level meaning-making, rather than a focus on the individuals, their psychology or their intent, we also implemented Hookway's (2008, p. 178) suggestion of 'moderate disguise', and do not name producers.

We watched the vlogs, and transcribed the audio text, retaining some gross-level detail around visual aspects. We focused primarily on textual meaning, utilising a critical-constructionist approach to thematic analysis (Braun & Clarke, 2006). Our analytic interest was focused 'outwards'. What story about veganism is being told in these vlogs? What 'truths' are being told? What 'lessons' might a viewer take away? Our engagement with the data involved initial coding across semantic (explicitly stated) and latent (implicit or more conceptual) meaning. We developed broad candidate themes initially centred on the 'lessons' from the videos, conceptualising the vlogs as instructional tools. Upon review, we shifted the analytical focus, resulting in three key themes pertaining to the construction of veganism. The themes were developed based on their central role in the story of veganism being told, and its rhetorical construction as a means of healthy eating (and wellness). The three themes were: 1) veganism for health and wellbeing; 2) veganism for abundance and pleasure; and 3) veganism as a learned skill.

Lessons from vegan vlogs

These vloggers fitted the dominant contemporary profile of the wellness celebrity – such as Deliciously Ella or Gwyneth Paltrow (Abouarrage, 2017): most were based in rich Anglo-Western countries (United States/Australia), and the vlogs were almost exclusively presented by young, white, able-bodied, slender Anglo-Western women (there was one solo man vlogger; one vlog was presented by a man and a woman). Our brief analysis captures the dominant ontologising practices: veganism in these vlogs was located within neoliberal healthist and contemporary privileged (white) wellness imperatives. Other constructions of veganism – as ethical eating from an animal rights perspective (Janssen, Busch, Rödiger, & Hamm, 2016) or as environmentally responsible eating (Greene-baum, 2017; Sandal, 2018) – were sometimes present (see also Kierans, 2017), but appeared only as minor characters in the story.

Veganism for health and wellbeing

The vlogs featured extensive implicit and explicit references to health. The central premise of this theme is that going vegan will be beneficial for improving your health. This played out in relation to three aspects: first, that veganism will improve your physical health through nutrition, and second, your mental health through resolving 'bad relationships' with food. Third, a vegan diet was also constructed as 'intuitively' healthy through what we refer to as body optimisation.

Physical health/mental health

References to physical and/or mental health ranged from specific aspects (e.g., energy or digestion) to looser articulations of 'feelings':

> I definitely noticed my digestive issues have gotten quite a bit better since giving up dairy and red meat and that kind of stuff.
>
> *(MLB)*

> This was a thirty-two-ounce glass of this super detoxifying amazing juice ... didn't taste the best but I can see that it would really help digestion get things going.
>
> *(SQ)*

> I'm vegan because I ... wanted to get healthy and I lost a lot of weight being vegan and I feel really great now.
>
> *(ML)*

The data contained numerous claims to improved physical health as a result of adopting a vegan diet. Claimed benefits tended to relate to improved bodily functions such as digestion, illness prevention and weight regulation. In this first quote, digestion is articulated within a before-and-after logic, a body problem (somewhat) resolved. In the second, the extreme health claims for a product are positioned loosely, and improving digestion claimed in the absence of any noted prior problem.

Another health benefit typically noted – as in the third extract above – was weight loss, although it tended not to be situated as the primary reason for being vegan, but rather as an (almost incidental) side effect. Body size and weight loss have been connected to better mental health for vegans (e.g., Beezhold, Radnitz, Rinne, & DiMatteo, 2015; Kierans, 2017), with weight loss almost an implicit proxy for improved health (within limits). In our data, eating vegan was positioned as significantly influencing wellbeing through improved mental health – through better 'relationships' with food, or eating, or the body (alongside weight loss, often). Veganism was constructed as (therefore) beneficial for mental health.

More than that, the dataset was flooded with the concept of 'feeling' (and feeling great, as in this third quote), meshing the physical, the psychological and the affective for wellbeing. Although others have connected the positive 'mentally healthy' aspects of veganism through the mechanism of morality, and an alignment between eating practice and a moral or good life (Kierans, 2017), in the vlogs this tended to be more loosely articulated.

Body optimisation

In a slightly different route to health and wellbeing, the subtheme body optimisation captured the recurrent claim that through veganism you will know your body and what it needs to reach its 'optimal' state, knowledge people were positioned as otherwise cut off from. This was present in the ideas that veganism offers your body a way to reach its optimal state, that it is essential to listen to your body (Sandal, 2018) (weight loss was often implicitly part of this). The body in these accounts has its own agency and its own 'wisdom' which we were constructed as otherwise ignoring. For instance:

> So personally what I'm eating today would be what I would eat in a day for a healthy body because I'm listening to my body I'm honouring my cravings I'm eating a majority of whole plant based foods that I know are going to nourish my health and make me whole and happy and healthy.
>
> *(CS)*

As in this quote, an idea of 'the truth of the body' played out across the dataset. The body was constructed as knowing intuitively what it needed for optimal wellness, with listening to it being key to achieving wellness. This 'truth' co-exists alongside another discourse around cravings – that they are wrong, and your head is what should guide food choices, not what you crave (evident in arguments against sugar, for instance). There, cravings tend to be situated as 'bad learnings', and in the vegan vlogs, as reflecting the truth of the body. In the dataset, the vegan person is positioned as having conquered this falsehood and reached a deeper truth where the body does not lie:

> You know food is just so it's just like this ... foundation of a good happy healthy life in your body ... remember that it's not just to feel full and it's not just to have ... to get abs or anything like that food is to feed your temple.
>
> *(CG)*

This quote evokes the body as a temple, a popular religious trope in healthist wellness discourse, which implicitly situates eating as a moral responsibility to avoid desecration. This evocation of the body as a 'holy site' evokes the 'Eastern mysticism' popular across much wellness discourse (Vandenburg & Braun, 2017) and aligns with an uncomfortable 'fetishisation' of 'ethnic' foods within some of these wellness diets and white food discourse, something Bailey (2007) has dubbed 'cultural food colonisation'. Such ideas construct a contemporary Western lifestyle and eating as contaminating and evoke a nostalgia for some (mythical) pure past.

Veganism as abundance and pleasure

Veganism was regularly constructed as pleasurable – connecting to a wider socio-cultural discourse around food, pleasure and living well (Finn, 2011). Sometimes pleasure was described and claimed explicitly, usually through reference to the deliciousness of the food being consumed. For instance:

> A huge heirloom tomato which I love they're my favourite this one's extremely delicious.
>
> *(KF)*

> This oatmeal literally tastes like an apple pie to me and if you guys have never tried baked oatmeal before it's so delicious it's like super thick and decadent.
>
> *(CS)*

Evidencing the rhetorical technique of 'extreme case formulations' (Pomerantz, 1986), vlogs often articulated hyperbolic claims such as 'extremely delicious', 'so delicious' and 'super thick and decadent'. Such language works to counter a construction of veganism as absence or deprivation of pleasure. Aesthetic pleasure was also connected to the visual presentation of food, a tension between 'what food looks like' and what it (might) taste like, that has been articulated in relation to social media, and its influence around food (Lee, 2017; Lewis, 2008; Sandal, 2018).

In a different way, in a normative context where eating is a gendered activity often characterised by restriction (for women), and in relation to a mode of eating that is inherently restrictive, we were fascinated by a theme of abundant eating, that seemed in and of itself to operate as a proxy for pleasure:

> Ok I don't know if you realize but this bowl is huge it's more like a serving bowl 'eh' but I'm probably gonna fill the whole thing up.
>
> *(CG)*

> He also kind of would always throw in like you know you don't have to restrict yourself on a vegan diet like you know it's all about eating large quantities and you know feeling full and satisfied after meals and I was like … this sounds like a dream like it's all I've ever wanted in life is to be able to eat until I'm full and not feel guilty and that's like it was so much about that for me at the beginning.
>
> *(BR)*

The notion of an abundant way of eating offered a broader and more conceptual idea of a positive, beneficial way of eating (see also Dejmanee, 2016). Vloggers referenced the amount they were eating (sometimes in claimed response to criticism or concern from viewers), constructing veganism in a way that leaves no room for eating 'pathology'. Indeed, as BR's quote illustrates, abundant eating in veganism interconnects with Theme 1 (health), particularly through mental health, and a release from restricted gendered eating expectations (Kierans, 2017).

However, there is a problematic tension here, because the freedom to eat abundantly seems deeply underpinned by a logic that this is 'possible' because a vegan diet means the person will not gain weight. This perhaps illustrates a 'false liberation' from body concerns (Dejmanee, 2016), and (neoliberal) imperatives towards (ultimate) health and wellness predominate.

Veganism as learned skill

Finally, veganism was constructed as a learned skill, a counter-normative way of eating that required a learning and an unlearning from previous practice and experience. Vloggers often positioned themselves as having become skilled in veganism – the skill being to ensure an informed and (claimed) nutritionally robust way of eating. Vegan eating required learning on the part of the vloggers – a narrative that predominates in contestations over how to do veganism 'well' (Hancox, 2018). In setting up their stories thus, the vloggers position themselves as 'everywomen', and vegan or plant-based eating – in a 'healthful' way – as open to all. At the same time, the skill of eating thus positions it as somehow exclusive, with the vloggers having learned skill and expertise. Overall, they seem to occupy a performative position somewhere between amateur and expert, or both:

> I've learned a lot about health and nutrition as I've gone through I think I didn't know a lot when I got started and I don't know if I was the healthiest person I was eating plant based but I dunno if was eating enough and since then I've learned a little bit more and I can 'um' … be a little bit more comfortable saying that I'm getting enough of everything 'um' however it's obviously a learning experience and I'm constantly learning more.
>
> *(TGGG)*

This theme aligns with other analyses of vegan vlogs where education (self and others) is constructed as an essential part of veganism (Kierans, 2017). The articulation of an ongoing learning journey also aligns with broader neoliberal logic to constantly work on the self. The task becomes one of ongoing effort and ongoing potential for improvement to be a better vegan person; sacrifice in eating has some moral connection to 'goodness' (Finn, 2011).

However, learning the truth of, and practising, vegan eating also required an unlearning, or a new learning that 'corrects' what had come before:

> It's really important to question everything that you've heard and that you think you know about food … I'm doing this challenge to … help people understand that they don't need to take protein powders they don't need to eat a bunch of animal products … in order to gain muscle and these industries these billion dollar industries and perhaps even trillion dollar industries that are built on the suffering of animals … and the illusion that people need to eat them and the things that come out of them … this is the

reason why I'm doing what I'm doing to show that everything that you've believed in up until this point is an illusion.

(GFT)

Here, vegans are positioned as having access to an authentic and true way of eating, once the 'wool' has been removed from their eyes. GFT evokes the 'red pill' of The Matrix, a now-popular meme for being beyond 'dupeism', with strong associations with misogynistic men's movements (Love, 2013). GFT also gently articulates an 'activist' vegan position – more radically expressed by (critiqued) 'vegan' documentaries such as 'Cowspiracy' and 'What the Health' – that pits plant-based eating against the bad world of Big Food. Veganism does not appear wholly or solely constructed around the individual, and the healthist pursuit of wellness, but the potency of this more political discourse was muted. Whether its potency will (again) increase as personal and planetary health become more widely connected (e.g., Willett et al., 2019), and with what implications, remains an empirical question.

Conclusion

We aimed to explore what veganism might look like to someone searching through some of the most popular social media modes. Veganism as currently constructed in the popular digital space of vlogs appears as a movement dominated by young, white, middle-class, slender women. This group of wellness (micro)celebrities/social media influencers both represent and construct what veganism 'looks' like, creating a capitalist (social media) space that veganism then occupies (Wright, 2015). Such online representations of veganism, part of the broader 'mediated biopolitics of contemporary foodscapes' (Goodman et al., 2017, p. 162), are likely highly influential to foodways, identities and reasons for why a person might adopt a vegan or plant-based diet.

Within these vlogs, veganism is situated not (primarily) within moral eating discourse, but within health and wellbeing discourse. Much of the discourse generated by dominant vegan figures in the digital space divorce it from its history as an animal-rights movement in favour of wellness discourses (Wright, 2015). Although animal rights and environmental concerns did feature in our data, the predominant focus was on the self and self-improvement – through evocations of physical state, mental state, and body size and embodiment. Vegan vlogs illustrate a way of eating outside/beyond current standard public health advice around food and nutrition (as exemplified in national healthy eating guidelines). Representationally, they align and intersect with other currently popular restrictive and claimed healthy eating styles, which are premised on very particular restrictions but rhetorically constructed as healthier and more enlightened than other modes of eating. As these constructions reflect and connect with increasingly pervasive postfeminist neoliberal self-improvement mandates (e.g., see Elias, Gill, & Scharff, 2017; Riley et al., 2018), we situate vegan vlogs as neoliberal identity projects, fitting healthist imperatives to work on, and better, the self.

Further reading

Greenebaum, J. (2012). Veganism, identity and the quest for authenticity. *Food, Culture & Society, 15*(1), 129–144.

Sampson, P. (2013). Contested frontier: Examining YouTube from a critical perspective. *E-Learning and Digital Media, 10*(3), 276–284.

References

Abouarrage, N. (2017, September 7). A guide to all of Gwyneth Paltrow's unique approaches to health and wellness. *W Magazine.* Retrieved from www.wmagazine.com/story/gwyneth-paltrow-health-wellness-quotes-goop

Bailey, C. (2007). We are what we eat: Feminist vegetarianism and the reproduction of racial identity. *Hypatia, 22*(2), 39–59.

Bandinelli, C., & Arvidsson, A. (2013). Brand yourself a changemaker! *Journal of Macromarketing, 33*(1), 67–71.

Beer, D. (2013). *Popular Culture and New Media: The Politics of Circulation.* Houndmills: Palgrave Macmillan.

Beezhold, B., Radnitz, C., Rinne, A., & DiMatteo, J. (2015). Vegans report less stress and anxiety than omnivores. *Nutritional Neuroscience, 18*(7), 289–296.

Berryman, R., & Kavka, M. (2017). 'I guess a lot of people see me as a big sister or a friend': The role of intimacy in the celebrification of beauty vloggers. *Journal of Gender Studies, 26*(3), 307–320.

Bertella, G. (2018). Vegetarian for a day or two. In A. Hardy, A. Bennett & B. Robards (Eds.), *Neo-tribes: Consumption, Leisure and Tourism* (pp. 33–49). Cham, Switzerland: Palgrave Macmillan.

Bratman, S. (2017). Orthorexia vs. theories of healthy eating. *Eating and Weight Disorders - Studies on Anorexia, Bulimia and Obesity, 22*(3), 381–385.

Braun, V., & Clarke, V. (2006). Using thematic analysis in psychology. *Qualitative Research in Psychology, 3*(2), 77–101.

Crawford, R. (1980). Healthism and the medicalization of everyday life. *International Journal of Health Services, 10*(3), 365–388.

Dejmanee, T. (2016). 'Food porn' as postfeminist play: digital femininity and the female body on food blogs. *Television & New Media, 17*(5), 429–448.

Delaney, M., & McCarthy, M.B. (2014). Saints, sinners and non-believers: The moral space of food. A qualitative exploration of beliefs and perspectives on healthy eating of Irish adults aged 50–70. *Appetite, 73*, 105–113.

Elias, A., Gill, R., & Scharff, C. (Eds.). (2017). *Aesthetic Labour: Beauty Politics in Neoliberalism.* London Springer.

Finn, S.M. (2011). *Aspirational eating: Class anxiety and the rise of food in popular culture* (PhD). University of Michigan. Retrieved from https://deepblue.lib.umich.edu/handle/2027.42/86292

Fox, N., & Ward, K. (2008). Health, ethics and environment: A qualitative study of vegetarian motivations. *Appetite, 50*(2), 422–429.

Goodman, M.K., Johnston, J., & Cairns, K. (2017). Food, media and space: The mediated biopolitics of eating. *Geoforum, 84*(Supplement C), 161–168.

Greenebaum, J. (2017). Questioning the concept of vegan privilege: A commentary. *Humanity & Society, 41*(3), 355–372.

Greenebaum, J., & Dexter, B. (2018). Vegan men and hybrid masculinity. *Journal of Gender Studies, 27*(6), 637–648.

Groopman, J. (2017, April 3). Is fat killing you or is sugar? What we do and don't know about dietary science. *The New Yorker*. Retrieved from www.newyorker.com/magazine/2017/04/03/is-fat-killing-you-or-is-sugar

Guardian, The. (2019, August 19). Our obesity crisis and its many different causes. Retrieved from www.theguardian.com/society/2018/aug/19/our-obesity-crisis-and-its-many-different-causes

Haenfler, R., Johnson, B., & Jones, E. (2012). Lifestyle movements: Exploring the intersection of lifestyle and social movements. *Social Movement Studies, 11*(1), 1–20.

Hancox, D. (2018, April 1). The unstoppable rise of veganism: How a fringe movement went mainstream, *The Guardian*. Retrieved from www.theguardian.com/lifeandstyle/2018/apr/01/vegans-are-coming-millennials-health-climate-change-animal-welfare

Harper, A.B. (2013). Going beyond the normative white 'post-racial' vegan epistemology. In P. Williams-Forson & C. Counihan (Eds.), *Taking Food Public: Redefining Foodways in a Changing World* (pp. 155–174). New York: Routledge.

Harrington, S., Collis, C., & Dedehayir, O. (2019). It's not (just) about the f-ckin'animals: How veganism is changing, and why that matters. In *Alternative Food Politics: From the Margins to the Mainstream* (pp. 135–150). London: Routledge.

Henderson, J.A., Ward, P.R., Coveney, J.D., & Taylor, A. (2009). 'Health is the number one thing we go for': Healthism, citizenship and food choice. Paper presented at the The Future of Sociology. Retrieved from https://tasa.org.au/wp-content/uploads/2015/03/Henderson-Julie.pdf

Hookway, N. (2008). 'Entering the blogosphere': Some strategies for using blogs in social research. *Qualitative Research, 8*(1), 91–113.

Hookway, N. (2017). Archives of everyday life: Using blogs in qualitative research. In V. Braun, V. Clarke & D. Gray (Eds.), *Collecting Qualitative Data: A Practical Guide to Textual, Media and Virtual Techniques* (pp. 166–188). Cambridge: Cambridge University Press.

Janssen, M., Busch, C., Rödiger, M., & Hamm, U. (2016). Motives of consumers following a vegan diet and their attitudes towards animal agriculture. *Appetite, 105*, 643–651.

Jenkins, H. (2014). Rethinking 'rethinking convergence/culture'. *Cultural Studies, 28*(2), 267–297.

Kaplan, A.M., & Haenlein, M. (2010). Users of the world, unite! The challenges and opportunities of Social Media. *Business Horizons, 53*(1), 59–68.

Khamis, S., Ang, L., & Welling, R. (2017). Self-branding, 'micro-celebrity' and the rise of Social Media Influencers. *Celebrity Studies, 8*(2), 191–208.

Kierans, K. (2017). *Common themes in the vegan YouTube movement: Community, health, ethics, regret and education* (Bachelor's thesis, Thompson Rivers University, British Columbia, Canada). Retrieved from https://core.ac.uk/download/pdf/84872074.pdf

Koch, F. (2017) *Recipe for Success: A Qualitative Investigation into the Role of Social Capital in the Gendered Food Blogosphere*. Media@ LSE Working Paper Series. London: LSE.

Lange, P.G. (2007). Publicly private and privately public: Social networking on YouTube. *Journal of Computer-Mediated Communication, 13*(1), 361–380.

Lee, S. (2017, December 23). Picture perfect? How Instagram changed the food we eat. *BBC News*. Retrieved from www.bbc.com/news/uk-england-london-42012732

Levinovitz, A. (2015). *The Gluten Lie: And Other Myths about What You Eat*. New York: Simon and Schuster.

Lewis, T. (2008). *Smart Living: Lifestyle Media and Popular Expertise*. New York: Peter Lang.

Lewis, T. (2018). Digital food: From paddock to platform. *Communication Research and Practice, 4*(3), 212–228.

Love, D. (2013, September 16). Inside red pill, the cult for men who don't understand women. *Business Insider*. Retrieved from www.businessinsider.com.au/the-red-pill-reddit-2013–8

Lupton, D. (2015). Health promotion in the digital era: A critical commentary. *Health Promotion International*, *30*(1), 174–183.

Lupton, D. (2017). Cooking, eating, uploading: Digital food cultures. In K. Lebesco & P. Naccarato (Eds.), *The Bloomsbury Handbook of Food and Popular Culture* (pp. 66–81). London: Bloomsbury Academic.

McCartney, M. (2016). Clean eating and the cult of healthism. *BMJ*, *354*, i4095.

Ministry of Health. (2013). *Eating for Healthy Adults*. Wellington: Ministry of Health.

Missbach, B., & Barthels, F. (2017). Orthorexia nervosa: Moving forward in the field. *Eating and Weight Disorders – Studies on Anorexia, Bulimia and Obesity*, *22*(1), 1.

Molyneaux, H., O'Donnell, S., Gibson, K., & Singer, J. (2008). Exploring the gender divide on YouTube: An analysis of the creation and reception of vlogs. *American Communication Journal*, *10*(1), 1–14.

Moreno, M.A., Goniu, N., Moreno, P.S., & Diekema, D. (2013). Ethics of social media research: Common concerns and practical considerations. *Cyberpsychology, Behavior, and Social Networking*, *16*(9), 708–713.

Mycek, M.K. (2018). Meatless meals and masculinity: How veg★ men explain their plant-based diets. *Food and Foodways*, *26*(3), 223–245.

NHS. (2016). *The eatwell guide*. Retrieved from www.nhs.uk/live-well/eat-well/the-eatwell-guide/

Phillipov, M. (2019). Introduction: Thinking with media: margins, mainstreams and the media politics of food. In M. Phillipov & K. Kirkwood (Eds.), *Alternative Food Politics: From the Margins to the Mainstream* (pp. 1–19). London: Routledge.

Pollan, M. (2008). *In Defense of Food: An Eater's Manifesto*. London: Penguin.

Pomerantz, A. (1986). Extreme case formulations: A way of legitimizing claims. *Human Studies*, *9*, 219–229.

Riley, S., Evans, A., & Robson, M. (2018). *Postfeminism and Health*: London: Routledge.

Ruby, M.B., & Heine, S.J. (2011). Meat, morals, and masculinity. *Appetite*, *56*(2), 447–450.

Samadi, D. (2017, January 6). Sugar is not only a drug but a poison too. *Huffington Post*. Retrieved from www.huffpost.com/entry/sugar-is-not-only-a-drug-but-a-poison-too_b_8918630

Sandal, C. (2018). *You are what you eat online: The Phenomenon of mediated eating practices and their underlying moral regimes in Swedish "What I eat in a day" vlogs*. (Master's thesis, Lund University, Sweden). Retrieved from http://lup.lub.lu.se/luur/download?func=downloadFile&recordOId=8943596&fileOId=8943599

Sassatelli, R. (2004). The political morality of food. Discourses, contestation and alternative consumption. In M. Harvey, A. McMeekin & A. Warde (Eds.), *Qualities of Food. Alternative Theoretical and Empirical Approaches* (pp. 178–191). Manchester: Manchester University Press.

Schneider, T., Eli, K., McLennan, A., Dolan, C., Lezaun, J., & Ulijaszek, S. (2019). Governance by campaign: The co-constitution of food issues, publics and expertise through new information and communication technologies. *Information, Communication & Society*, *22*(2), 172–192.

Senft, T.M. (2008). *Camgirls: Celebrity and Community in the Age of Social Networks*. New York: Peter Lang.

Song, F.W. (2010). Theorizing Web 2.0. *Information, Communication & Society*, *13*(2), 249–275.

Tandoh, R. (2016, 14 May). The unhealthy truth behind 'wellness' and 'clean eating'. *Vice*. Retrieved from www.vice.com/en_au/article/jm5nvp/ruby-tandoh-eat-clean-wellness

Vandenburg, T., & Braun, V. (2017). 'Basically, it's sorcery for your vagina': Unpacking Western representations of vaginal steaming. *Culture, Health & Sexuality*, *19*(4), 470–485.

Vegan Society, The. (nd). Definition of veganism. Retrieved from www.vegansociety.com/go-vegan/definition-veganism

Willett, W., Rockström, J., Loken, B., Springmann, M., Lang, T., Vermeulen, S., … Murray, C.J.L. (2019). Food in the Anthropocene: The EAT-Lancet Commission on healthy diets from sustainable food systems. *The Lancet*, *393*(10170), 447–492.

Williams, S.N., & Nestle, M. (Eds.). (2018). *Big Food: Critical Perspectives on the Global Growth of the Food and Beverage Industry*. London: Taylor & Francis.

Wright, L. (2015). *The Vegan Studies Project: Food, Animals, and Gender in the Age of Terror*. Athens, GA: University of Georgia Press.

PART 3
Expertise and influencers

7

A SEAT AT THE TABLE

Amateur restaurant review bloggers and the gastronomic field

Morag Kobez

Introduction

By the end of the twentieth century, food had become big business. It began to be positioned as a source of entertainment in late modern capitalist societies, with myriad television food shows, food magazines, celebrity chefs and food festivals combining to create a 'food culture' (Gargano, 2014). For those who consider eating out a form of entertainment or a lifestyle – so-called 'foodies' – the life-sustaining function of food is subordinate to those other functions. Popular use of the term 'foodie' to describe such people arose from use of the word in *The Official Foodie Handbook* by Barr and Levy (1984). De Solier (2013a), Getz, Robinson, Andersson and Vujicic (2014), and Cairns, Johnston and Baumann (2010) emphasise personal identity-formation as being central to foodies. De Solier (2013b, p. 9) defines foodies as people for whom food is central to a sense of self: 'amateur enthusiasts who strive to form a moral self not only through the consumption of material cultures of food, but also their production'. De Solier (2018) also observes that the practice of photographing meals and sharing images via social media has become mainstream. More than 326 million posts tagged '#food' on Instagram in 2019 support this assertion (Instagram.com).

The proliferation of social and digital media in recent decades has allowed 'serious leisure' amateur food bloggers to become a prominent feature and subset of this foodie culture. Serious leisure is defined as 'the systematic pursuit of an amateur, hobbyist or volunteer' core life activity that its adherents find so interesting, meaningful and fulfilling that 'they launch themselves on a (leisure) career centred on acquiring and expressing a combination of its special skills, knowledge and experience' (Stebbins, 2007, p. 5). Digital media have enabled such amateur foodies to construct public identities around consumption of restaurant experiences, and to participate extensively in the discourse around restaurant dining.

This knowledgeable amateur participation is now encroaching on the discursive territory previously occupied by a small number of elite professional food critics.

In this chapter, drawing on interviews with prominent food bloggers undertaken in 2017, I analyse the changes brought to the gastronomic field of culture by the rise of digital media. I argue that the activities and outputs of such bloggers challenge the primacy of professional restaurant critics as cultural intermediaries in the gastronomic field. This study draws on my experience of the restaurant-reviewing worlds. I have been a professional restaurant reviewer for the past eight years, an amateur blogger for four years prior to that, and someone who has paid close attention to online consumer reviews (OCRs) for more than a decade. The changes I observed in the practice of restaurant reviewing and the critical discourse around cuisine were the impetus for this research. Since I began writing professional reviews in 2010, I detected tensions between professional critics and the growing numbers of amateur food bloggers and online contributors in the foodscape.

My previous research has shown how the proliferation of amateur restaurant reviews on digital media has changed the practices of the professional food critic – including changes to the process of carrying out reviews, the ethical framework guiding this process and the format of reviews (Kobez, 2018). The present study seeks to determine whether amateur food bloggers can be considered as cultural intermediaries in the twenty-first-century gastronomic field and competitors to professional food critics in this role. While some work has been done on the relationship between OCRs and professional reviewing, my research is the first to investigate the transformative nature of online and digital media in the gastronomic field at the micro-level of professional restaurant reviews and their digital competitors in the form of serious leisure amateur bloggers.

The chapter begins by establishing gastronomy as a cultural field, with critical reviewers as cultural intermediaries playing an important constitutive role. It details the ethical framework that professional reviewers have used to guide their conduct and the writing of restaurant reviews in order to establish and maintain a disinterested orientation from commercial concerns and personal preferences. This is followed by analysis of empirical data gathered from interviews with food bloggers according to the themes of ethics, authenticity and authority. The chapter concludes with a summary of key findings indicating a more contested gastronomic field with a hierarchy of cultural intermediaries.

Cultural intermediaries and the gastronomic field

Bourdieu introduced the concept of 'field' to denote and delineate 'the state of a cultural enterprise when the relevant production and consumption activities achieve a certain degree of independence from direct external constraints' (Ferguson, 2004, p. 104). For the agents operating within a field, Bourdieu's concept of 'habitus' explains how the discourse that constitutes a field, and its norms and standards of value, 'become embodied and internalized in the cognitive structure of agents' (Husu, 2013, p. 266). Ferguson (2004, pp. 104–106) argues that from around the middle of the eighteenth century in France, gastronomy developed into a cultural enterprise, exhibiting embodied norms and standards, and sufficient independence from direct external constraints so as to be considered a cultural field. Gastronomy refers to the evaluation of culinary experiences rendered primarily in text.

Entering into abstract aesthetic discourse about the consumption of cultural products and experiences, mediated by experts, is the hallmark of critical discourse, and is co-constitutive of the cultural field in question. Bourdieu (1984, p. 325) introduced the term 'new cultural intermediaries' in his major work, *Distinction: A Social Critique of the Judgement of Taste*. He defines cultural intermediaries as 'the producers of cultural programmes on TV and radio or the critics of "quality" newspapers and magazines and all the writer-journalists and journalist-writers', who have assigned themselves the role of 'divulging legitimate culture' (Bourdieu, 1984, p. 326). In any field of cultural production – art, literature, film, gastronomy – the critical review is constitutive of the discourse that establishes the referents and standards by which judgements of taste and differentiations of quality can be made and accepted as legitimate. A discourse is understood here as a set of meanings, representations, images and statements of how people represent themselves and their social world (Holttinen, 2014, pp. 575–576).

An orientation of 'disinterestedness' from commercial concerns and direct personal preference is a necessary condition to produce critical discourse. In the 1950s, an ethical framework for restaurant reviewing was established by Craig Claiborne, the pioneering food critic for *The New York Times* (Blank, 2006; Kapner, 1996; Sietsema, 2010). This framework became institutionalised as professional practice among restaurant reviewers in the latter decades of the twentieth century. Claiborne's four-part ethical framework consists of reviewer anonymity; allowing restaurants a grace period after opening before a review is undertaken; the publication (the reviewer's employer) paying for all dining; and the production of a long-form review article of 500–700 words (Voss & Speere, 2013). Davis (2009) suggests that reviewer anonymity originates from an identified need for disinterestedness – meaning distance and detachment from the object or performance under consideration, in order to be able to make accurate and legitimate aesthetic judgements. Adherence to this ethical and procedural framework allowed professional reviewers to maintain a detachment from considerations other than the intrinsic quality of the restaurant experience and its relation to the aesthetic values of the gastronomic field.

Cultural fields and the discourses which produce and sustain them are sites of social struggles over status and power (Shrum, 1996, p. 11). Demarcating the boundaries of a field, its players and criteria for value and distinction, and the legitimacy to make or influence such classifications, is subject to continuous contestation (Bourdieu, 1993, p. 137). It stands to reason that as new amateur agents enabled by social and digital media seek to enter cultural fields, the existing cultural intermediaries will be displaced, leading to disruption and an upheaval of the field (Bourdieu, 1984, p. 231). This provides a compelling theoretical basis from which to explore how the penetration of amateur reviewers into the gastronomic field in the early twenty-first century has changed the nature and composition of the field.

Online and digital media has made available the technical means for amateurs to publish their own reviews and engage directly in critical discourse. However, not all amateurs display the necessary disinterestedness from personal preferences. Whereas restaurant-reviewing blogs can often be situated within the realm of serious leisure, with specialised skills and knowledge acquired over time, consumer review sites such as Yelp.com and Zomato are, according to Rousseu (2012, pp. 60–61), more representative of 'everyman'. I claim that only serious leisure food bloggers can be considered as competitors to professional critics as cultural intermediaries. This is because reviews on OCR sites focus overwhelmingly on personal preferences and do not generally engage with abstract discourses that are the hallmark of the critical reviewer.

Comparative analyses of the content of user-generated (blogs, OCRs) and professional reviews are under-researched. There is only one article at the time of writing that directly engages with the aims of this chapter, and then only in part. Vásquez and Chik (2015, p. 231) argue that 'online restaurant reviews provide a means through which individuals can display their culinary capital … as they establish their expertise on matters such as authenticity, taste, quality, and the perceived value of their dining experiences'. Their research seeks to broaden the cultural context of studies of OCR sites by comparing 120 user-generated reviews of one-star Michelin restaurants in Hong Kong on OpenRice, with those in New York City on Yelp.com. They argue that 'online reviews … highlight a paradox central to the … growing 'democratizing tendency' in contemporary food discourses' (Vásquez & Chik, 2015, p. 232). On the one hand, OCR sites 'provide an alternative to elite forms of restaurant reviewing, in allowing for the presence of – and access to – a greater diversity of voices, perspectives and opinions about 'high end' dining than ever before'. On the other hand, they observe that reviews of Michelin-starred restaurants on these platforms may 'simultaneously reproduce existing social hierarchies' (Vásquez & Chik, 2015, p. 232).

However, the discussion of the results of the authors' analysis, and the illustrative quotes provided, indicate that the user-generated reviews in this sample were overwhelmingly descriptive in terms of taste, ingredients, aroma, portion size and value for money, with little evidence of abstraction to wider aesthetic or ethical discourses. Where this was employed, writers invoked tradition and heritage in using simple descriptors such as 'authentic Korean food' or traditional 'regional cuisine', without further elaboration. Culinary 'capital' was displayed by reference to famous restaurants and celebrity chefs. However, rather than abstracting their experience to broader aesthetic discourses, the OCR writers sought to create culinary capital by referring to their own narrative of dining at Michelin-starred restaurants. The evidence presented suggests that writers of

OCRs reduce the culinary experience to one of personal narrative and preferences, and therefore cannot be considered as agents or cultural intermediaries in the gastronomic field.

In their 2007 study, Johnston and Baumann analysed culinary-focused articles in all 2004 issues of four major US print-format magazines: *Bon Appetit*, *Saveur*, *Food and Wine* and *Gourmet*. In contrast to OCRs, the authors found that professional reviewers consistently deployed 'authenticity' as their dominant discursive strategy. Authenticity is an elusive quality – touching on aesthetics, ethics and emotions – hinging upon an often-ambiguous set of expectations held by a knowing audience. Once familiar with a cultural field, a member can quickly assess the authenticity of its artefacts. The term is used in many contexts to describe a range of desirable traits including credibility, originality, sincerity, naturalness, genuineness, innateness, purity or 'realness' (Grazian, 2010, p. 191). Different cultural fields have their own signifiers of authenticity, which will be readily understood by those immersed in that habitus (Johnston & Baumann, 2015, p. 63).

'Authenticity', as deployed by professional reviewers, is operationalised in Johnston and Baumann's (2007, pp. 179–180) US study through the frames of geographic specificity, simplicity (referring to handmade, non-industrial, and/or organic or naturally raised produce), personal connection (with the produce and ingredients), traditional connection with food (referring to food produced according to traditional, rustic methods), and exoticism considered as unusualness (rarity or atypicality) and foreignness. The authors undertake a content analysis, specifically a frequency count, of these frames within the magazine articles (Johnston & Baumann, 2007, pp. 176–178). At first glance, the tendency for high-status food to include casual dining experiences drawn from a range of rustic, simple 'authentic' fare again seems to refute Bourdieu's theory of distinction. However, as Johnston and Baumann (2007, p. 197) find, elements of Bourdieu's analysis remain crucial to understanding the legitimation of status and cultural distinction in a democratic age. Covert status markers in food writing balance the need for distinction with democratic equality (Johnston & Baumann, 2007, p. 197). These boundaries are demarcated by these frames denoting 'authenticity'. The ability to recognise and articulate aesthetic criteria and elicit them from a culinary experience is a crucial element of how professional reviews seek to create distinction through authenticity.

In 2014, I replicated Johnston and Baumann's (2007) study, applying their frames to conduct a discourse and content analysis of Australian restaurant reviews in four mainstream media print-version food guides (Kobez, 2016). In extending their study, I also analysed OCRs of the same restaurants. The resulting data affirmed that similar themes reflecting authenticity are also consistently identifiable in Australian professional restaurant reviews (Kobez, 2016). However, amateur restaurant reviews on OCR sites do not use these frames. Rather, they use frames which are overwhelmingly concerned with direct personal preferences and pre-conceived, everyday expectations, including the size of meals, the price of the food, the service or conduct of restaurant staff and the restaurant's décor. These attributes are often raised in reference to the personal event occasioning the meal such as a birthday or anniversary (Kobez, 2016). Amateur reviews on OCR sites are thus often structured, as Bissell (2011, p. 154) has also pointed out, around the incongruence between diners' anticipated expectations versus their perceptions of the actual restaurant experience.

Professional restaurant reviews continue to articulate an abstract discourse of taste and distinction, which is constitutive of any cultural field. Reviews on OCR sites do not, nor do they seek to do so, being primarily consumer-oriented and concerned with direct personal preferences. There are no existing studies that address whether and how online amateur participants, such as the serious leisure food blogger, can be considered as cultural intermediaries in the gastronomic field.

Methodology

The present study builds on my recent work which investigated how the proliferation of digital media platforms has disrupted and transformed the tenets of traditional restaurant criticism established by Claiborne (Kobez, 2018). This research identified a number of key changes to professionals' reviewing practices: the requirement for immediacy in reviewing restaurants (replacing the customary 'grace period' to allow restaurants to establish themselves); the loss of professional critics' anonymity resulting from the need to establish a public profile in competition with online and digital media platforms; a move away from long-form, critical reviewing to the short format of listicles; and an erosion of the detachment and integrity that allows for 'objective' evaluation and negative criticism. I found that professional restaurant critics have been forced to accommodate the new pressures and formats brought by the contemporary digital media environment. This change is primarily due to the commercial pressures on their mainstream media employers in competition with online and digital media platforms. I concluded that the digital disruption and transformation of the mainstream media industry has eroded the 'disinterestedness' of the traditional restaurant reviewer, which is the foundation for the production of critical discourse (Kobez, 2018).

This chapter builds on this work to understand how amateur food bloggers interpret their role and practices, and their relationship to the professionals who operate within the gastronomic field. Primary data for the study was sourced from interviews with author-owners of 10 restaurant reviewing blogs. Eligible bloggers met the following criteria: they were based in the three largest Australian cities of Sydney, Melbourne and Brisbane, and had generated a sustained and substantial body of work (with regular restaurant reviews posted for a minimum of 1.5 years with at least a total of 50 restaurant reviews). The sample of blogs consisted of *Australian Foodie* (Brisbane), *All About Food* (Sydney), *Little Miss Melbourne* (Melbourne), *Foodie Mookie* (Sydney), *Eat Drink + be Kerry* (Brisbane), *Simon Food Favourites* (Sydney), *A Young Gourmet* (Brisbane), *I'm so Hungree* (Melbourne), *The Pursuit of Food Perfection* (Sydney) and *Espresso &*

Matcha (Brisbane). The Queensland University of Technology Human Research Ethics Committee approved the study. All respondents agreed to be identified by the name of their blog.

Participants were first emailed via their blog contact details requesting their participation in an interview. Once provided with the project information and informed consent to participate, suitable times were agreed upon, and interviews were conducted by telephone or in person. Open-ended interviews with semi-structured questions were conducted in the second half of 2017. The duration of these interviews was 45–60 minutes. Respondents were asked about the reasons they started their blog, and also their reasons for continuing to operate a blog over a sustained period of time. I also enquired as to how much time they spent writing and curating their blog and whether they received any payment via advertising, or payment in kind in the form of meals and invitations to events. I sought to understand their motivation for writing reviews. Questions were also asked about the process and format of the reviews, and if their blogging practice incorporated an aesthetic dimension and ethical framework. I encouraged participants to expand and elaborate on their comments throughout the interviews, and mention anything further that arose from the various topics discussed.

In analysing the interview transcripts, my research design followed the 'grounded theory' approach, that 'through open coding, theoretical sampling and constant comparison answers will emerge' (Grbich, 2013, p. 81). A line-by-line analysis of the transcripts of interviews with food bloggers was first undertaken to identify preliminary themes and allocate codes to them. As multiple passes through the data were undertaken, these codes were refined to the concepts of ethics, authenticity and authority.

Ethics

Professional critics often perceive that the penetration of food blogging into their traditional sphere of expertise had eroded, and in some areas completely dismantled, the ethical framework for restaurant reviewing established by Claiborne and outlined above (Kobez, 2018). One of the most important findings of the present study is that the negativity of professional reviewers regarding the 'interestedness' of amateurs was almost completely unfounded in the sample of serious leisure amateurs that were interviewed. This negativity is based on the perceived susceptibility of amateurs to the public relations 'hosting model' where members of the media are invited to visit restaurants for free, with the understanding that they will feature the venue in their work, usually in a positive way (Kent, 2008; Waddington, 2012). By contrast, all bloggers interviewed employed an implicit ethical framework in their practice, emphasising respect, fairness, honesty, representativeness of dishes at restaurants and disclosure of invitations or free meals – consistent with their disinterestedness from commercial motivations. All amateur respondents indicated that their blogging practice was intrinsically motivated by interest, enjoyment and 'passion', rather than driven by commercial or employment considerations. No respondent received an income from their activity and few incorporated advertising on their blog.

Most of the amateur bloggers interviewed were conversant with Claiborne's ethical framework without necessarily being aware of its author or origins. Most of the respondents reported that they have an implicit ethical framework by which they operate. For example, on this point, *Espresso & Matcha* (interview, 29 October 2017) said, 'I do have a pretty clear code of ethics, although I don't have anything formal about this on my blog.' In addition to not accepting or disclosing free meals and invitations, bloggers' ethical frameworks emphasised respect for the restaurateur, representativeness in choices of dishes, and sufficient and accurate information to make an informed evaluation, as demonstrated below:

> Ideally there are four people because then I will wait to see what everyone chooses to see what interests people and I will order something different. Ideally everyone orders a different dish so there's a better idea of the menu. There are instances where we eat the entire menu in one sitting. The other thing is I will always ask for the signature dish and order that. I'll also ask if there's a special as that gives you an indication of what might be fresh in that day.
>
> *(Foodie Mookie, interview, 7 March 2017)*

> It's hard being stuck between that old-school mentality of journalistic ethics and being in the newer social media movement, but in my position I do have things I will and won't do and it's not formal or written anywhere, but I don't take payment for posts, I try to have as much background information before I write things. There's an informal one.
>
> *(I'm so Hungree, interview, 20 January 2017)*

Similarly, for The Pursuit of Food Perfection (interview, 16 May 2017):

> I have a personal code of ethics ... To always be truthful, be respectful, I mean pretty basic stuff. Being respectful of the people who run the restaurant, and not necessarily not saying something's bad, but rather than being quick to jump to the conclusion that it's terrible, take a step backwards to think about the reason why it's bad.

This evidence is representative of amateur respondents, who all stated that they operate according to an implicit ethical framework, even though most did not have this stated explicitly on their blog. Where they accepted invitations, every respondent without exception indicated that they disclosed to readers if a meal was invited or paid for by the restaurant. This disinterestedness from commercial concerns, which is the hallmark of the amateur, allows serious leisure bloggers to engage with the aesthetic and ethical discourses characteristic of the cultural intermediary.

Authenticity

A cultural field, as defined in this chapter, is co-constituted by the abstract discourse of critics, who operate as intermediaries between the creators or producers of cultural artefacts and knowledgeable audiences for them. In consuming artistic performances or cultural products, the discourse of critics moves from direct, personal experience to higher levels of abstraction – in terms of wider aesthetic, ethical, historical, cultural or geographic contexts. As elaborated above, professional restaurant reviewers deploy abstract frames related to authenticity in their writing in order to demonstrate their status as arbiters of taste in a more democratic age of cultural 'omnivorousness' where traditional high–low culture distinctions have lost much of their resonance (Peterson, 1997).

However, the analysis of blogger transcripts found that where amateurs engaged in specific aesthetic discourses regarding authenticity (considered as provenance of food, personal connection with produce, traditional methods in its preparation, and foreignness or 'exoticism'), it was almost always linked to an ethical framework of environmental sustainability. For example, *Australian Foodie* (interview, 9 May 2017) said:

> Yes, definitely I will write about it, because I like eating local produce. I think that's important environmentally and from a local economic point of view, and the more those things are normalised in reviews then the more people will take notice of that. Or I'll ask the question, where is this beef from and they say, overseas, then that gives them something to think about.

This orientation is different from the professional critic, where my research indicates that aesthetic engagement with the culinary experience is intended more as an end in itself. By contrast, most of the amateurs interviewed linked these aesthetic categories to a more concrete ethical consideration or outcome. This is demonstrated in the examples below, when respondents were asked about whether their reviews highlighted 'authenticity'. In this context, *Foodie Mookie* (interview, 7 March 2017) linked provenance of food with minimising its 'carbon footprint':

> One of the things I'm very passionate about is understanding the footprint of where I'm dining. I've done quite a bit of travel around food experiences and one of the joys is going to a location and they're using local beef from local farmers and really minimising the carbon footprint and giving you a snapshot of what that area produces in terms of food.

Similarly, regarding provenance and personal connection, *Australian Foodie* (interview, 9 May 2017) was concerned with 'food miles':

> Handmade? Yes, definitely again that makes a difference it's a personal touch to what they're making, and it also hopefully means they're using more raw ingredients rather than processed stuff. [It] feeds back into the

environmental food miles and the other things like preservatives and stuff that people don't want to eat.

The term 'food miles' refers to food sourced by major retailers from global supply chains that travels long distances before consumption, often by air or road, thus consuming large quantities of fossil-fuel and contributing to climate change (Smith et al., 2005).

Using predominantly seasonal, local produce rather than global agricultural supply chains was emphasised by *I'm so Hungree* (interview, 20 January 2017):

> To me personally it's important [provenance] and … if it's a feature of the menu I will definitely raise it. It is important. I get really annoyed if I go to a café and they're serving asparagus out of season. That's not ok with me. I don't want to see those tiny skinny asparagus from Peru. That really gets me. I do believe in seasonality that things shouldn't be coming super-far.

The linking of authenticity to an ethic of environmental sustainability, local production and personal connection denotes a resistance to generic products sourced through the complex supply chains of globalised agriculture. This orientation is somewhat different from the professional critic, where my research suggests that aesthetic engagement with foreignness and exoticism – skinny asparagus from Peru as in the above example – is considered as an end in itself regardless of food miles or other environmental considerations.

Serious leisure amateur bloggers do engage with abstract aesthetic and ethical discourses, and thus can be considered cultural intermediaries in the gastronomic field akin to their professional counterparts. However, compared with professionals they are less strategic in their deployment of abstract frames. In reflecting on their restaurant-reviewing practices, where amateurs did invoke an aesthetic discourse in a holistic, experiential sense, the language used was vague and not clearly defined – tending to use terms such as 'atmosphere', 'vibe' and 'ambience'. They only rarely linked their critical discourse to other aesthetic or geographical contexts, unlike professionals, who commonly refer to overseas trends and experiences, discourses and publications.

Authority

The amateur bloggers interviewed for this study all demonstrated significant expertise in culinary discourses and gastronomic experiences. All respondents indicated that they invested serious effort in their leisure activity, often dining out multiple times per week and produced a number of substantive weekly posts, in addition to other undertakings, as illustrated below:

> I do spend a lot of time writing my posts … I will usually write at least a paragraph or two about each dish, as well … what made me visit or

what I was expecting, so that adds up to usually around 300–500 words a post, and a typical post will have 10 or so photos – the interior décor, and sometimes the exterior, the menu, a picture of each dish, maybe the table setting, and maybe drinks. So it adds up, and it's not unusual for me to spend 10–14 hours a week on the blog. Between my fulltime job and the time spent actually eating out, then up to 14 hours documenting it – you get the idea I don't have much time left to do other stuff. Other than my paid job, this is what I spend the bulk of my time doing – eating out and blogging about it.

(Espresso & Matcha, *interview, 29 October 2017*)

Once in general over the year I try to get to all the new opened major restaurants in Sydney ... It can be almost full-time if I'm travelling, going out and eating, then writing and eating again, it's all consuming and ... [doesn't] feel like a holiday.

(Foodie Mookie, *interview, 7 March 2017*)

One respondent reported that they had been doing unpaid work experience in restaurants to enhance their culinary expertise:

With blogging I'm always keeping up to date with all the news around the Brisbane food scene through the paper and through online as well and also looking at reviews on places that have been open a while that I haven't got around to trying. I've started doing work experience in restaurants so I can improve my food knowledge so I can have more of an opinion on it ... I think I will try and be more critical as a result of having this experience.

(A Young Gourmet, *interview, 25 January 2017*)

In choosing which restaurants to review, some respondents indicated that they engaged in deeper critical analysis of mainstream reviews to identify gaps, deficiencies or where more information might be useful for their readers. In this, amateur bloggers display the spirit and approach of academic enquiry in their practice. This is consistent with their tertiary education, but also indicates expertise as demonstrated below:

I think it's always good to get an idea about stuff they've talked about and perhaps what they've missed, and how I could possibly improve on it and I think that's a good way, if another place has covered it, I like to have a read, see where I can improve and ask questions ... that they didn't touch on.

(A Young Gourmet, *interview, 25 January 2017*)

I try to visit everything I think will be good, or all the main places in the cuisines I like – Japanese and Italian. In terms of writing, for me it often comes down to I want to add something. I want to add a useful opinion and not just write for the sake of writing. I hate that. So every piece I write, I try and pass some knowledge or review insight about the food itself, or the cuisine I'm talking about.

(The Pursuit of Food Perfection, *interview, 16 May 2017*)

In this sense, bloggers engage directly with the work of professionals in a critical manner to identify gaps in the discourse, or where another 'angle' or perspective might be employed. They often reviewed the same restaurants or linked to where a review of the same restaurant has been previously undertaken on their blog, and often spoke of professionals in a way that suggested a respect for professionals' authority. This evidence suggests that rather than operating in parallel, serious leisure amateurs directly engage with the work of professionals, and in many cases, review the same restaurants.

Despite this direct critical engagement, another important finding was that most bloggers continued to take their major cues, in terms of new trends and culinary experiences, from the professional critics in mainstream media publications. Mainstream media sources were still considered authoritative in terms of major themes, trends and aesthetics. *A Young Gourmet* (interview, 25 January 2017) for example, expected that professionals would be more knowledgeable: 'you've got to expect professional food writers to have more knowledge about the food and the restaurant. There's always going to be a place for blogs and user-generated reviews, but myself I prefer to read professionals.'

Foodie Mookie (interview, 7 March 2017) referred to a leading mainstream publication as a 'bible', indicating its authoritative status:

I am an extensive reader as well as writer, so I love the mainstream media and have great respect for traditional food reviewers. *Gourmet Traveller* is a bit of a bible in terms of what's coming up and what's established. I've used the guide for ticking off three, two and one hatted restaurants [highly rated restaurants that have been awarded 'Chef's Hats' by the Australian Good Food Guide Awards] to visit and review, which has been a bit of fun.

(Foodie Mookie, *interview, 7 March 2017*)

I'm so Hungree (interview, 20 January 2017), *Eat, Drink + be Kerry* (interview, 1 February 2017) and *The Pursuit of Food Perfection* (interview, 16 May 2017) all indicated that they view mainstream publications as authoritative because of their research and expertise, with the *Good Food Guide* and *Gourmet Traveller* mentioned repeatedly in this regard by amateur respondents.

This evidence indicates that while serious leisure bloggers may be considered agents in the gastronomic field, they consider themselves as subordinate to professional critics as cultural intermediaries. In most cases, they accepted this as legitimate, and did not question that professional critics should be held to higher standards in their knowledge of the culinary experience and expertise in their writing about it, and consequently, to be assigned a higher status in the gastronomic field.

One amateur respondent noted that professional critics guard these positions jealously (*Eat, Drink + be Kerry*, interview, 1 February 2017), which is consistent with Bourdieu's field theory – that well-socialised actors will instinctively seek to defend their position and habitus against the entry of new competitors.

As a cultural intermediary, the role of the reviewer is co-constitutive of any artistic or cultural field, as it is the critic that generates the discourse about the cultural product or experience, which feeds back into the creative process of the producers of that experience. In order for serious leisure amateurs to be competitors to professional reviewers as cultural intermediaries, they must, by definition, demonstrate a disinterested, abstract orientation to their discourse, moving beyond the discussion of everyday personal preferences that dominates OCRs. This evidence generated for this study indicates that serious leisure amateur bloggers do engage with abstract aesthetic and ethical discourses, and thus can be considered cultural intermediaries in the gastronomic field. While serious leisure amateurs demonstrated abstract aesthetic and ethical discourses in their practice, their orientation tended more towards the ethical, and their aesthetic discourse was less purposefully deployed, and less sophisticated in its articulation compared with professional critics.

Conclusion

Serious leisure amateur food bloggers, who may be considered as a subset of foodie culture, are now agents and cultural intermediaries in the gastronomic field, as they produce aesthetic and ethical discourses regarding their culinary experiences. This status is due to their disinterestedness from commercial considerations, the implicit ethical standards they employ, and the aesthetic and ethical criteria with which they engage. However, amateurs continue to take their cues from professional critics in terms of new trends, themes, events and establishments.

I conclude that there remains a hierarchy within the gastronomic field where professionals are afforded higher status by bloggers and other audiences. However, the gastronomic field is now larger and more contested with diffuse boundaries, rather than being the rarefied domain of a small number of elite, professional restaurant reviewers. In this it marks a profound transformation of the gastronomic field brought by the proliferation of online and digital media.

Further reading

De Solier, I., & Duruz, J. (2013). Food cultures: Introduction. *Cultural Studies Review*, *19*(1), 4–8.

Grove, D.J. (2007). Global cultural fragmentation: A Bourdieuan perspective. *Globalizations*, *4*(2), 157–169.

Hyde, Z. (nd). Omnivorous gentrification: Restaurant reviews and neighborhood change in the downtown eastside of Vancouver. *City & Community*, *13*(4), 341–359.

Johnston, J., & Baumann, S. (2007). Democracy versus distinction: A study of omnivorousness in gourmet food writing. *American Journal of Sociology*, *113*(1), 165–204.

References

Barr, A., & Levy, P. (1984). *The Official Foodie Handbook*. London: Elbury Press.

Bissell, D. (2011). Mobile testimony in the Information Age: The power of travel reviews. *International Journal of Cultural Studies, 15*(2), 149–164.

Blank, G. (2006). *Critics, Ratings, and Society: The Sociology of Reviews*. Lanham, MD: Rowman & Littlefield.

Bourdieu, P. (1984). *Distinction: A Social Critique of the Judgement of Taste*. Trans. R. Nice. Cambridge, MA: Harvard University Press.

Bourdieu, P. (1993). *The Field of Cultural Production: Essays on Art and Literature*. Cambridge: Polity Press.

Cairns, K., Johnston J., & Baumann, S. (2010). Caring about food: Doing gender in the foodie kitchen. *Gender & Society, 24*(5), 591–615.

Davis, M. (2009). *A Taste for New York: Restaurant Reviews, Food Discourse, and the Field of Gastronomy in America* (unpublished PhD dissertation, New York University).

De Solier, I. (2013a). *Food and the Self: Consumption, Production and Material Culture*. New York: Bloomsbury.

De Solier, I. (2013b). Making the self in a material world: Food and moralities of consumption. *Cultural Studies Review, 19*(1), 9–27.

De Solier, I. (2018). Tasting the digital: New food media. In K. LeBesco & P. Naccarato (Eds.), *The Bloomsbury Handbook of Food and Popular Culture* (pp. 54–65). New York: Bloomsbury.

Ferguson, P.P. (2004). *Accounting for Taste: The Triumph of French Cuisine*. Chicago and London: Chicago University Press.

Gargano, S. (2014). IBISWorld Australian Industry Reports. Retrieved from http://clients1.ibisworld.com.au.ezp01.library.qut.edu.au/reports/au/industry/ataglance.aspx?entid=2010

Getz, D., Robinson, R., Andersson, T., & Vujicic, S. (2014). *Foodies and Food Tourism*. Oxford: Goodfellow.

Grazian, D. (2010). Demystifying authenticity in the sociology of culture. In J.R. Hall, L. Grindstaff and M-C. Lo (Eds.), *Handbook of Cultural Sociology* (pp. 191–200). New York: Routledge.

Grbich, C. (2013). *Qualitative Data Analysis: An Introduction*. Thousand Oaks, CA: Sage.

Holttinen, H. (2014). How practices inform the materialization of cultural ideals in mundane consumption. *Consumption Markets & Culture, 17*(6), 573–594.

Husu, H.-M. (2013). Bourdieu and social movements: Considering identity movements in terms of field, capital and habitus. *Social Movement Studies, 12*(3): 264–279.

Instagram [hashtag food]. (2019, 15 March). Retrieved from www.instagram.com/explore/tags/food/

Johnston, J., & Baumann, S. (2010). *Foodies: Democracy and Distinction in the Gourmet Foodscape*. New York and London: Routledge.

Johnston, J., & Baumann, S. (2015). *Foodies: Democracy and Distinction in the Gourmet Foodscape* (2nd ed.). New York and London: Routledge.

Kapner, S. (1996). Craig Claiborne. *Nation's Restaurant News*, February, 66.

Kent, M. (2008). Critical analysis of blogging in public relations. *Public Relations Review, 34*(1), 32–40.

Kobez, M. (2016). The illusion of democracy in online consumer restaurant reviews. *International Journal of E-Politics, 7*(1), 54–65.

Kobez, M. (2018). Restaurant reviews aren't what they used to be: Digital disruption and the transformation of the role of the food critic. *Communication Research and Practice, 4*(3), 261–276.

Peterson, R.A. (1997). The rise and fall of highbrow snobbery as a status marker. *Poetics*, *25*, 75–92.

Rousseau, S. (2012). *Food and Social Media: You Are What You Tweet*. Lanham, MD: AltaMira/Rowman & Littlefield.

Shrum, Jr., W.M. (1996). *Fringe and Fortune: The Role of Critics in High and Popular Art*. Princeton, NJ: Princeton University Press.

Sietsema, R. (2010). Everyone eats… but that doesn't make you a restaurant critic. *Columbia Journalism Review*, 2 February. Retrieved from www.cjr.org/feature/everyone_eats.php?page=all

Smith, A., Watkiss, P., Tweddle, G., McKinnon, A., Browne, M., Hunt, A., … & Cross, S. (2005). *The validity of food miles as an indicator of sustainable development-final report*. Retrieved from www.national-academies.org

Stebbins, R.A. (2007). *Serious Leisure: A Perspective for Our Time*. New Brunswick, NJ and London: Transaction.

Vásquez, C., & Chik, A. (2015). 'I am not a foodie …': Culinary capital in online reviews of Michelin restaurants. *Food and Foodways*, *23*, 231–250.

Voss, K., & Speere, L. (2013). Food fight: Accusations of press agentry, a case for ethics and the development of the association of food journalists. *Gastronomica*, *13*(2), 41–50.

Waddington, S. (2012). *Tools and Tips for Working with Bloggers*, December 2. Retreived from https://wadds.co.uk/blog/2012/12/02/tools-and-tips-for-working-with-bloggers

8

I SEE YOUR EXPERTISE AND RAISE YOU MINE

Social media foodscapes and the rise of the celebrity chef

Pia Rowe and Ellen Grady

Introduction

In recent years, we have witnessed the rise of a new type of a celebrity: the celebrity chef. Many popular chefs, from the UK's Jamie Oliver to America's Martha Stewart and Australia's Donna Hay, have become household names in the developed world. In early television cooking shows, chefs demonstrated their cooking skills. Now, the celebrity chef is a performer, an entertainer and a lifestyle consultant (Barnes, 2014; Chiaro, 2008; De Solier, 2005; Lane & Fisher, 2015; Salkin, 2013). Where once these chefs may have gained a public profile from their cookbooks or television programs, now blogs and various social media channels provide opportunities for promoting their ideas beyond their culinary identities.

In effect, these chefs embody a wider range of food-based identities and are increasingly influential in defining what constitutes 'good food' in the contemporary foodscape, often telling consumers what they should be eating to define their identities, maintain their health and control their household budget (Johnston & Goodman, 2015). Given the influence food celebrities wield over consumers, concerns have been raised about the nature of their advice, with some describing it as 'data-free celebrity science' (Robbins, 2013; Rousseau, 2015). For Rousseau (2015), the problem is not simply that celebrities sometimes challenge existing scientific knowledge on what constitutes good, healthy food, but also that they tend to present an alternative with absolute certainty in a manner which completely disregards the very process of scientific inquiry.

Controversial Australian media personality Pete Evans, a former pizza chef turned paleo diet advocate, is a prime example of this new development. Evans originally gained popularity as a 'traditional' chef. Between 1998 and 2011 he opened numerous award-winning restaurants in Australia. He also hosted a range of television shows, including *Fresh with the Australian Women's Weekly* (2007–2009), and appeared on *MasterChef Australia*. Since 2010, Evans has been the judge on the popular reality show *My Kitchen Rules*, which won a Logie Award in 2014 for the 'Most Popular Reality Program' (*The Sydney Morning Herald*, 2014), and in 2015 was the highest rated reality television competition in Australia with approximately two million weekly viewers, thus cementing himself as a household name in Australia. In the US, Evans also hosted the show *A Moveable Feast*, which gained a 'Daytime Emmy Award' nomination in 2014 (Styles, 2014).

More recently, Evans has put considerable focus on his personal endeavours as an expert on the paleo diet, which requires people to cut out grains, legumes and most dairy products (Healthline, 2018). Since his transformation from a traditional chef to an activist, Evans has sensationalised the benefits of the diet as both the key to good health and a treatment for numerous health disorders and illnesses, including autism, asthma, diabetes and even cancer (Willis, 2018). He actively utilises his heavily moderated Facebook page to both market his enterprise and criticise those who disagree with his views.

In 2015, Evans attracted global mainstream media attention with his controversial co-authored cookbook, *Bubba Yum Yum: The Paleo Way for New Mums, Babies and Toddlers*. Several health experts raised alarms about the potential of some recipes to severely harm infants' health and the book was subsequently dropped by its publisher (*The Australian*, 2015; Davey, 2015; Taylor, 2015). Evans vehemently rejected all the experts' assessments and went on to self-publish the book.

Developments such as these raise crucial questions regarding contemporary food cultures and the dissemination of food-related information. How do famous celebrity chefs themselves use social media sites to construct their credibility and gain social and political influence? Does expertise on social media translate into mainstream media coverage? Given the fundamental importance of media in both shaping and reflecting the food cultures that constitute foodscapes, critical evaluation of both conventional and new media must be a core feature of such an analysis (Johnston & Goodman, 2015).

We respond to these questions through a close investigation of the mainstream print media coverage available online, focusing on the most popular media outlets in Australia as identified by Nielsen (2017), and the individual Facebook site of Pete Evans in the 12-month period surrounding the digital and hard copy releases (published on May 2015 and November 2015 respectively) of *Bubba Yum Yum*. By doing so, we explore the role of celebrity influencers in the construction of digital foodscapes more broadly.

Contemporary foodscapes and celebrity chefs

> And today's most intelligent quote must go to … the DAA (DIEtitians [*sic*] a$$ociation [*sic*] of Australia. These stalwarts of the nations [*sic*] health seem to be in love with the ingredient called … BREAD and their advice is really helping our nation become the sickest on the planet.
>
> *(Pete Evans, 1 August 2017, Facebook)*

Food has always been a site of contestation, and the increasing digitisation of information-sharing practices has created new opportunities for more varied forms of food activism. Whereas 'Alternative Food Networks' more broadly emerged in response to the large-scale, industrial systems as activists sought to gain greater social control of food provisioning, and to reframe notions of equity and fairness in trade relations (Goodman, DuPuis, & Goodman, 2012), digital food activism extends beyond immediate concerns with the food system, encompassing activist projects where food is a means to broader political action (Eli, Schneider, Dolan, & Ulijaszek, 2018).

As Eli et al. (2018) note, digital food activism does not simply refer to practices that occur on digital media. Rather, it encompasses forms of activism enabled and shaped through digital media platforms. In this context, key questions to consider include 'Who sets the agenda?', 'What values guide the agenda?', 'What evidence counts and who has expertise to provide it?' and 'What modes of action are employed and through which constituencies?' (Eli et al., 2018).

Lyson (2014) argues that in the US, the alternative food movement has become a 'master frame' for addressing a number of pressing social issues, including those related to public health, such as obesity. As Giordano, Luise and Arvidsson (2018) discuss, these new kinds of communities focusing on consumer goods or practices offer opportunities to marginalised people for social and political action. The community here, however, is not a starting point, but the 'imagined end-result of social action' (Giardano et al., 2018, p. 620). Yet, despite their potential to engage with everyday people, alternative food movements have been criticised for being in the domain of middle- and upper-class people (DuPuis & Gillon, 2009; Guthman, 2007; Lyson, 2014). Such criticism has obvious resonance with some of the new forms of food activism – such as that of celebrity chefs.

In recent years, food celebrities have gained tremendous cultural influence (Johnston, Rodney, & Chong, 2014). Johnston and Goodman (2015) posit that media, particularly in the guise of food celebrities, represent a fundamental component of contemporary 'foodscapes'. 'Foodscapes' are urban food environments in which consumers acquire, prepare, talk about and gather meaning from food. The concept covers both the cultural (values, meanings and representations) and the material (physical landscapes, ecologies and political economy) dimensions of food, while highlighting the dialectical relationship between food culture and food materiality (Johnston & Goodman, 2015). In this context, Johnston and Goodman (2015, p. 209) assert that, through its 'elevated celebrities, interactive processes and multiple media platforms', food media 'has both an empirical and a normative-aspirational dimension in that it frames and mediates what foodscapes could and should be, how they could and should operate, and for whom they could and should work'.

In a similar vein, Barnes (2014, p. 170) defines celebrity chefs as 'talking labels', arguing that they act as both a cultural intermediary and a boundary object to construct normative knowledge around choosing and shopping, as well as cooking and eating, while connecting audiences to food and themselves.

The issues and objects that food celebrities produce and consume are icons that communicate social norms, stereotypes and aspirations for the public (Johnston et al., 2014). In addition, food celebrities' 'messages' now often include the promise to make people 'better' in multiple ways: 'better cooks, better socially, better at caring for friends and family, better lifestyles and wellbeing, better homes, better connected to food and those producing it, better global citizens even' (Barnes, 2014, 170).

Much of the existing research links these developments to neoliberal food governance, which shifts the responsibilities of good citizenship from institutions to individuals (Guthman, 2007, 2008; Lewis, 2010; Lewis & Potter, 2010; Rose, 1996). In this approach, celebrity chefs reframe personal health as a privatised issue, so people's individual lifestyle choices become important sites of self-governance (Johnston & Goodman, 2015; Lewis, 2010). Lewis (2010, p. 587) argues that the celebrity lifestyle experts take this process even further by 'embodying and enacting models of consumer citizenship through their own much publicized and idealized domestic and personal lifestyles, which are played out across their various personae as experts, celebrities and private selves'. As such, the messages they convey extend beyond self-governance, as the idea of personal empowerment is simultaneously deeply embedded in the rhetoric (Johnston & Goodman, 2015). In addition, by leaving the definition of 'healthy' to each user's interpretation, people themselves can decide what constitutes a healthy lifestyle (Chung, Agapie, Schroeder, Mishra, Fogarty, & Munson, 2017).

However, celebrity chefs do not focus only on individual consumers. Indeed, Barnes (2014, p. 171) sees celebrity chefs as key actors in 'a new form of mediated food governance that seek to influence public relationships with food as well as policy making and food's wider politics'. Successful chefs not only influence laypeople, but also experts such as policymakers and scientists (Barnes, 2014; Eden, 2011; Johnston & Goodman, 2015). Famous individuals are powerful agents for spreading nutrition information, but problems arise when they spread health-related misinformation via media as the general public is prone to believe it (Hoffman & Tan, 2013; Myrick & Erlichman, 2019). The key concern here is that, over time, enabled particularly by the rapid expansion of social media, celebrity chefs have become experts in their own right rather than relaying information from other expert sources (Barnes, 2014). Given the lack of transparency about their qualifications, and their ability to reach large audiences through social media, which by its nature does not require fact checking, there is a risk that the messages they convey will have serious negative health consequences for consumers.

The role of social media in blurring the line between celebrities and traditional experts

> The idea of a lone consultant becoming, in three short years, more influential than entire university departments of Ph.Ds., is indicative of a new level of potential for celebrity in health messaging.
>
> *(Khamis, Ang, & Welling, 2017, p. 204)*

Traditionally, experts and celebrities have been thought of as existing in markedly different spheres of public life, which are linked to very different sets of values and logics (Lewis, 2010). Whereas experts and expertise were seen as synonymous with 'modes of rational knowledge and techniques of social organization that accompanied the rise of the modern state', celebrities were typically seen as 'co-extensive with popular and consumer culture, with a mediatized public sphere where entertainment is privileged over information, affect over meaning' (Lewis, 2010, p. 581). Rousseau (2015) argues that, in the realm of lifestyle media, the concept of expertise has been radically reconfigured along more popular lines, so the relationship between the celebrity and the expert is no longer marked by hierarchies, or a distinction between popular and consumer culture and the professional/governmental realm.

The increased media coverage of celebrity chefs has played a key role in their emergence as new contemporary food experts (Barnes, 2014). The fact that they are continuously on television screens, social media and bestseller lists gives celebrity chefs ample opportunities to place themselves, and be placed by the media, as powerful players (Barnes, 2014; Goodman, 2010; Guthman, 2007, 2008). New media allow regular people to broadcast their everyday food practices (De Solier, 2018). This importation of daily social experiences into the online domain across different social media platforms has further blurred the boundary between the private and the public, in turn both contributing to, and coinciding with the birth of the new lifestyle expert. De Solier (2018, p. 63) discusses the rise of the 'amateur foodie expert' and their new food media in relation to traditional media – first in the form of bloggers (versus professional journalists), and more recently YouTube cooks (versus traditional TV chefs). The focus of our chapter is on slightly different – though related – development: the evolution of a 'traditional chef' with an established brand and visibility on 'old media' to an amateur nutrition expert on a new media platform.

Recent research confirms that celebrities indeed function as a significant driving force behind the new diet-based communities online. Ramachandran et al. (2018) assessed the most 'liked' Facebook pages in Australia that made recommendations on healthy eating for their alignment with the Australian Guideline to Healthy Eating (AGHE). Of the nine pages analysed, only two fully aligned with government guidelines, while the rest deviated from AGHE in some way – either through direct contradiction, misinformation, or overly restrictive recommendations (Ramachandran et al., 2018). In addition, the four most popular pages (Michelle Bridges 12 Week Body Transformation; Jamie Oliver; Pete Evans; and I Quit Sugar) were hosted by celebrities. The implications of this research are clear. Facebook is a powerful tool for disseminating information, but currently the most popular content skews towards non-traditional experts, echoing Rousseau's (2015) sentiments of 'data-free celebrity science'.

This new transformation of celebrities into experts in their own right has been a global phenomenon. However, this is far from a homogenous space. The UK's Jamie Oliver, for instance, exemplifies a more conventional trajectory from a career as a traditional chef, to subsequently moving on to television – first with the purpose of demonstrating cooking to audiences, but gradually politicising his mission to influence healthy eating habits. In 2005 Oliver launched the *Feed Me Better* campaign aimed at changing the food eaten by school children and was subsequently voted as the 'Most Inspiring Political Figure of 2005' (BBC News, 2006). Oliver has since moved on to spearhead numerous other food activism campaigns internationally, including *Ministry of Food*.

While Oliver operates 'with' the existing systems of governance, Vani Hari, aka 'The Food Babe' in the US, illustrates another type of development in the food influencer arena. Hari describes herself as a 'revolutionary food activist' even though by her own admission, she has no qualifications in human nutrition (https://foodbabe.com/about-me/). In 2011 Hari started her blog, www.FoodBabe.com, based on her experiences of becoming sick as a result of eating a 'typical American diet'. Hari's main mission is to challenge 'big business' and enable her audience to 'experience a richer sense of health' by 'assimilating the information' on FoodBabe.com. She frequently mobilises her followers, whom she refers to as 'The Food Babe Army' to target big companies, and lists numerous 'accomplishments' on her website as the companies respond to the consumer pressure. The accuracy of Hari's health claims and fears have been widely challenged by scientists such as the analytical chemist Yvette d'Entremont (2015), yet she has established a significant following (as of 20 March 2019, Hari had over 1.2 million 'likes' on Facebook [Hari, 2019]) and authored best-selling books of health advice.

Goodman, Johnston and Cairns (2017) argue that to understand this production of food knowledge we need to look at the role and creation of truth discourses, and the truth claims at the heart of contemporary foodscapes. One significant concern here is the fact that, at present, there are no regulations about the claims concerning food and nutrition that celebrity chefs can make. Social media enable celebrities to control the message, while their audience may be entirely made up of fans who choose to follow them (Johns & English, 2016). They can present food-based information at odds with the information given by institutions regulated by government. Consequently, when arguably powerful celebrities such as Pete Evans or Vani Hari present information as facts, they are not held to the same standards of truth/evidence as official bodies such as the Dieticians Association of Australia (DAA), or the Food and Drug Administration (FDA) in the US.

Of course, these 'truth discourses' are not a totalising force, but rather sites of frequent contestation and conflict (Goodman et al., 2017). Barnes (2014) found that the different ways in which audiences talk back to chefs can also create moments of resistance. However, we must note here that not all social media are created equal, as all platforms have their unique logics and rules for interaction. While followers on platforms such as Twitter can respond to 'bad' information, and act as a 'swift correctional resource' (Rousseau, 2012b), on other platforms, such as Facebook, the moderator of the page can control who gets to participate in the discussion, thus creating an echo chamber that is harder to penetrate.

Media and technology have been key factors in shifting perceptions and attitudes about nutrition during the last 50 years (Vaterlaus, Patten, Roche, & Young, 2015). Research has shown that around 80 per cent of all internet users aged 18–46 go online for health information (Lohse, 2013) and social influencers are a primary factor in the adoption of health behaviours (Centola, 2013). More specifically, work on eating disorders suggests that social media contribute to an echo-chamber effect, where people believe their values are more common than they actually are, because they selectively view contributions by similarly minded people (Turner & Lefevre, 2017). The prospective wide reach of potentially dangerous health information through online channels has seen professional nutritionists demand 'best practice' industry guidelines to ensure quality control, but the nature of social media currently defies such containment (Tobey & Manore, in Khamis et al., 2017).

Information overload is also crucial, with consumers increasingly bombarded with messages about what to eat from multiple sources. In the social media landscape, audiences' attention is scarce and short-lived and boundaries between public and private, professional and amateur and newsworthy and Instagram-worthy constantly shift; so, there is extraordinary pressure for the content to be attention-grabbing (Rousseau, 2015). Indeed, in a world where the boundaries between personal and public are blurring at a dizzying speed, it is easy to imagine a follower (mis)interpreting information presented on social media as a method approved for public practice (Rousseau, 2012b). The problem here is that the need to stand out may contribute to the circulation of hyperbole and misinformation, thus further confusing audiences about which sources are trustworthy. The most pessimistic accounts of this 'attention economy', where expertise becomes conferred with celebrity status and popularity, describe the phenomenon as the 'death of expertise' (Collins, 2014; Nicholas, 2014; Rousseau, 2015).

At the same time, acquiring the status of an expert and influencing everyday food practices is not only contingent on media coverage, but also on the ability of chefs to appear accessible, 'real' and authentic to their audiences, downplaying their celebrity status (Barnes, 2014; Johnston & Goodman, 2015). The role of social media in creating online communities plays a crucial role here, as 'food activists often try to exploit the community dimension to affect the market dimension' (Sinischalchi & Counihan, 2014, p. 10). Social media enable celebrities to present themselves in a more personal way, increasing connection with the audience, and socialising with like-minded individuals provides opportunities for open communication, in turn contributing to feelings of trust.

However, while traditional television chefs may have been used for brand endorsement, the celebrities themselves now are the brand they are selling to audiences (Johns & English, 2016). As such, food celebrities now embody both cultural and material dimensions by being the key voices of cultural and culinary authority, but also, simultaneously, being embedded in the political economies of media production, food marketing and creation (Johnston & Goodman, 2015).

Celebrity chef Pete Evans: a case study

Methods

Some 13.3 million Australians use online channels to source information on news and current affairs (Nielsen, 2017). Recognising the significantly contrasting views of accredited health experts and Evans in the public health debate regarding his paleo cookbook for infants, *Bubba Yum Yum*, we used media framing content analysis (Matthes & Kohring, 2008) to understand how Evans sought to establish his status as an expert in public wellbeing in an attempt to further this cause, while his reputation was being brought into question by various health experts in the mainstream media. While this particular approach has been criticised for not providing quantifiable results (Matthes & Kohring, 2008), our aim here is to use Evans as an illustrative example, and thus we make no claims regarding the generalisability of the analysis.

We focused on the most popular media outlets in Australia as identified by Nielsen (2017): news.com.au; nine.com.au; ABC News Websites; smh.com. au; *Daily Mail Australia*; *The Guardian*; Yahoo 7 News Websites; BBC; *Herald Sun*; and *The Age*. We collected articles published in these media outlets over a 12-month period, subdivided in two periods: the three-month period leading up to the scheduled release of *Bubba Yum Yum* by Australia's Pan Macmillan (15 December 2014 to 15 March 2015); and the nine-month period following Pan Macmillan's public announcement that it would no longer be publishing the book (16 March 2015 to 15 December 2015). The articles were found using the Lexis Nexis database searching three terms, 'Bubba Yum Yum', 'Pete Evans' and 'Paleo'. We also analysed Pete Evans' Facebook feed over the same time-period. Here, no search parameters other than the timeframe were used. Only the initial post made by the Facebook page administrator was included in the analysis, as our focus was on Evans' communication rather than that of his followers.

In total, the search terms returned 30 articles for the first time-period, and 104 digital media articles for the second period. A total of 711 posts were made by an owner or administrator of Pete Evans' Facebook page between 15 December 2014 and 15 December 2015. The reports generated in Lexis Nexis for both the mainstream media coverage and Evans' Facebook data were uploaded to NVivo, and a word-frequency query was run to establish key themes. Based on the most frequently occurring words on Evans' Facebook page, we then conducted further keyword searches on the mainstream media coverage to compare the two framings.

The case of Bubba Yum Yum

After Pete Evans changed his life to live *The Paleo Way*, he argued that food can be medicine; indeed, it should be our first port of call for a healthier life (Evans, 2015b). A former internationally renowned chef, restaurateur and television personality, Evans now markets himself as a certified health coach, with qualifications from the Institute for Integrative Nutrition (IIN), who wants to change the lives of everyone around him (Evans, 2015b). IIN was founded in 1992 and offers an online 'Health Coaching Training Program' for fee-paying students. IIN is licensed by the New York State Education Department (IIN, nd), but not by the Accreditation Council for Education in Nutrition and Dietetics in the US (Academy of Nutrition and Dietetics, 2019).

As of March 2019, Pete Evans had 1,515,741 followers on Facebook, 207,000 followers on Instagram and 19,906 subscribers on YouTube. While these numbers are small compared to some other celebrity food figures (Jamie Oliver had 6.8 million Facebook followers as of March 2019), Evans nonetheless has potential to influence multiple social groups in this media space, as such warranting careful scrutiny of the narratives he creates around 'healthy food'. Critically, Goodman et al. (2017, p. 164) describe Evans' foodscape brand as offering diagnoses of, and solutions to, food-related social problems, presenting consumers with idealised food choices, chastising some market actors, while celebrating others.

A recent example of the conflict between different 'expertise' was evident in Evans' cookbook *Bubba Yum Yum*. The book contained recipes based upon a strict paleo diet, including one which captured the attention of journalists, while horrifying health experts. Evans et al. offered bone broth for babies as a substitute to commercial infant supplement formula. Professor Heather Yeatman, the then-President of the Public Health Association of Australia, argued that there was a very real possibility that an infant could die if the book was published, noting that the recipe contained a potentially toxic amount of vitamin A for infants (*The Australian*, 2015). The public backlash from practising health experts and lay publics caused a strong response, eventually leading to the publisher, Pan Macmillan, cancelling the title's publication (Quinn & McNab, 2015).

In response, Evans self-published the book, both online and in hard copy, arguing that many individuals were seeking its 'beautiful treasure trove of nutritional recipes' (Quinn & McNab, 2015). He asserted that the recipe in question had been in print for years, with no recorded cases of harm (*The Guardian*, 2015). While Evans subsequently altered the recipe, reducing the vitamin A content, he insisted the publishers were overreacting because of the negative public response, rather than out of any real concern for babies' health (Australia's Channel 7 Sunday Night news, in *The Morning Bulletin*, 2015). Given the significant potential risks to public health involved in this case, Evans' claims to expertise requires analysis.

'A simple guy, who knows his stuff': opportunity or threat to new food cultures?

In the same way alternative food networks originally emerged as a response to the large-scale industrial food systems as activists sought to gain greater social control of food provisioning (Goodman et al., 2012, the rise of the social media food influencer can be seen as a critique of current institutions of food knowledge. What is interesting here, however, is that the social media influencers such as Pete Evans do not necessarily direct their critique at the market. Rather, they can also target the government, media, or the research and science that inform government policies. Consider the following examples:

> Critics don't really seem to grasp what paleo is. It's not a diet, nor is it nutritionally incompetent nor is it a fad. All these labels serve to do is disconnect paleo far away from its reality. At its heart, paleo is a way of life and a balanced approach to returning to eat the most natural foods available, straight from the source.
>
> *(Evans cited in Tran, 2015)*

> I would also like to extend a huge thank you to the DAA [Dieticians Association of Australia], the misinformed media and everyone else that has helped promote this book in their own special way! With your help *Bubba Yum Yum* has been consistently in the top selling books in ibooks since it was released earlier this year, and inspired me to make it available globally;)
>
> *(Evans, 2015a)*

Even though Evans positions himself as the expert relaying the information, there is a significant and observable focus on building the community throughout Evans' Facebook posts. In particular, his use of the word 'tribe' when he talks to, and about, his audience is telling. His neoliberal emphasis on individuals, rather than institutions, as the authorities on good health is consistent with his broader notion of everyday people as the subject matter experts.

> Wow – I am truly humbled by the responses to yesterdays [*sic*] post. Everyone is so passionate about the present and the future and this is why we have worked so hard to help build this community (Tribe) on this page.
>
> You are the leaders, the change makers, the people who will create a stronger future for so many people, and Nic and I are truly honoured to be apart [*sic*] of this movement with you all!
>
> The first step is through education, and growing a community of like minded [*sic*] individuals who are empowered and impassioned to stand up and make a change no matter how big or how small, knowing that each day they have the power to improve not only their own lives, but that of their families and of their community and I am so excited by what lies ahead from reading your stories that you continue to share on here.
>
> *(Evans, 2014)*

While a proper social class-based analysis is outside the scope of this chapter, two things ought to be emphasised here. First, while Evans utilises the word 'tribe' frequently when he talks to, and about, his audience, there is no doubt that this is an elite community. The material aspects required to live up to the expectations of Evans' prescribed diet are significant, and only accessible by those with adequate means and time to prepare the 'natural' diet. Furthermore, while this rhetoric creates an illusion of 'us', there is no escaping the fact Evans uses the community to sell his own program, *The Paleo Way* and other products such as his cookbooks.

Second, in the broadest sense, Evans may be contextualised as a marginalised actor creating a new space for food politics. However, a deeper scrutiny reveals that rather than democratising the space for food consumers, the digital food community created by Evans leaves no room for meaningful dialogue at the micro-level. Evans outright rejects any other experts' opinions, and deletes any dissenting comments and blocks the users from further commenting on his Facebook page (another Facebook group called 'Blocked by Pete Evans' was formed specifically by and for users who had been booted off his page). As such, he, as the influencer, purposefully amplifies the echo-chamber effect of a Facebook community.

While the mainstream media frame expertise almost exclusively in terms of institutionalised knowledge and use people with recognised qualifications as their expert sources, the clear disconnect between public health authorities and some Facebook audiences does raise significant concerns. However, it would also be elitist to assume that Evans' audiences blindly buy into his claims – and indeed the fact that so many have been blocked from his page indicates that this is far from the case. What it does signal, however, is that more stringent regulation on social media such as Facebook is required to better regulate the quality of nutritional and other health information offered by influencers such as Evans. In addition, it offers learning opportunities for public institutions for creating engaging content on their own platforms, and more effectively harnessing influencer power to spread public health messages.

Conclusion

The rapid expansion of social media has supported the rise of celebrity chefs, giving them platforms to share and promote their individual causes/interests. Media-savvy celebrity chefs may benefit from this publicity, even when the mainstream media coverage of their cause and activism is negative. As we observed, while Evans has successfully built his own social media following on Facebook, his influence is limited to his own followers. However, this does not mean that individual social media influencers' messages are insignificant. The fact that Evans can build such an extensive following, while resisting the fact checking of the mainstream media, is in itself a cause for concern. The bigger question here, however, is what implications this has on digital food cultures more broadly. The alternative space for politics created here offers a moment of resistance for the participating individuals. The irony, of course, is that while AFNs were often created in response to the market, influencers such as Evans may now use the alternative status to create a space for their own market enterprises. Finally, while it is not possible to make definitive statements regarding the actual public impact of individual influencers on food- and health-related matters, it is clear that more needs to be done to evaluate and regulate the information that influential people such as Evans can disseminate.

Further reading

Fraser, A. (2017). *Global Foodscapes: Oppression and Resistance in the Life of Food*. New York: Routledge.

References

Academy of Nutrition and Dietetics. (2019). *Accreditation Council for Education in Nutrition and Dietetics*. Retrieved from www.eatrightpro.org/acend

Australian, The. (2015, March 12). Pete Evans paleo cookbook for kids, Bubba Yum Yum, 'shelved'. Retrieved from www.theaustralian.com.au/life/food-wine/pete-evans-paleo-cookbook-for-kids-bubba-yum-yum-shelved/newsstory/33f9a4675e897ef71f5c02496b7a0d9a

Barnes, C. (2014). Mediating good food and moments of possibility with Jamie Oliver: Problematising celebrity chefs as talking labels. *Geoforum, 84*(1), 169–178.

BBC News UK. (2006). *Oliver wins ch 4 political award*. Retrieved from http://news.bbc.co.uk/2/hi/uk_news/politics/4673028.stm

Centola, D. (2013). Social media and the science of health behaviour. *Circulation, 127*(21), 2135–2144.

Chiaro, D. (2008). A taste of otherness eating and thinking globally. *European Journal of English Studies, 12*(2), 195–209.

Chung, C.F., Agapie, E., Schroeder, J., Mishra, S., Fogarty, J. & Munson, S.A. (2017). When personal tracking becomes social: Examining the use of Instagram for healthy eating. *Proceedings of the 2017 CHI Conference on Human Factors in Computing Systems*. ACM, Denver, Colorado, 6–11 May.

Collins, H. (2014). *Are We All Scientific Experts Now?* Cambridge: Polity Press.

Davey, M. (2015). *Pete Evans paleo for kids cookbook put on hold amid health concerns*. Retrieved from www.theguardian.com/australia-news/2015/mar/12/pete-evans-paleo-for-kids-cookbook-put-on-hold-amid-health-concerns

De Solier, I. (2005). TV dinners. Culinary television, education and distinction. *Continuum: Journal of Media & Cultural Studies, 19*(4), 465–481.

De Solier, I. (2018). Tasting the digital new food media. In K. LeBesco & P. Naccarato (Eds.), *The Bloomsbury Handbook of Food and Popular Culture* (pp. 54–65). London: Bloomsbury.

DuPuis, E.M., & Gillon, S. (2009). Alternative modes of governance: Organic as civic engagement. *Agriculture and Human Values, 26*(1–2), 43–56.

Eden, S. (2011). Food labels as boundary objects: How consumers makes sense of organic and functional foods. *Public Understanding of Science, 20*(2), 179–194.

Eli, K., Schneider, T., Dolan, C., & Ulijaszek, S. (2018). Digital food activism: Values, expertise and modes of action. In K. Eli, T. Schneider, C. Dolan & S. Ulijaszek (Eds.), *Digital Food Activism* (pp. 770–799). London: Routledge.

d'Entremont, Y. (2015). *The 'Food Babe' blogger is full of shit*. Retrieved from https://gawker.com/the-food-babe-blogger-is-full-of-shit-1694902226

Evans, P. (2014, December 3). [Facebook]. Retrieved from www.facebook.com/paleochefpeteevans/photos/wow-i-am-truly-humbled/747047425388646/

Evans, P. (2015a, November 1). [Facebook]. Retrieved from www.facebook.com/paleochefpeteevans/photos/a.170871629672898/919431004816953/?type=3&comment_id=919622908131096&reply_comment_id=919860834773970&comment_tracking=%7B%22tn%22%3A%22R%22%7D

Evans, P. (2015b). The Paleo Way website. Retrieved from https://thepaleoway.com/our-experts/#pete-evans

Giordano, A., Luise, V., & Arvidsson, A. (2018). The coming community. The politics of alternative food networks in Southern Italy. *Journal of Marketing Management*, *34*(7–8), 620–638.

Goodman, D., DuPuis, E.M., & Goodman, M.K. (2012). *Alternative Food Networks: Knowledge, Practice, and Politics*. Abingdon: Routledge.

Goodman, M.K. (2010). The mirror of consumption: Celebratization, developmental consumption and the shifting cultural politics of fair trade. *Geoforum*, *41*(1), 104–116.

Goodman, M.K., Johnston, J., & Cairns, K. (2017). Food, media and space: The mediated biopolitics of eating. *Geoforum*, *74*, 161–168.

Guardian, The. (2015, August 17). Paleo diet: Pete Evans says controversial baby broth never hurt anyone. Retrieved from www.theguardian.com/lifeandstyle/2015/aug/17/paleo-diet-pete-evans-controversial-baby-broth-never-hurt-anyone

Guthman, J. (2007). The Polanyian way? Voluntary food labels as neoliberal governance. *Antipode*, *39*(3), 456–478.

Guthman, J. (2008). Thinking inside the neoliberal box: The micro-politics of agrofood philanthropy, *Geoforum*, *39*(3), 1241–1253.

Hari, V. (2019). *About Vani Hari*. Retrieved from https://foodbabe.com/about-me/

Healthline (2018). *The paleo diet – A beginner's guide plus meal plan*. Retrieved from www.healthline.com/nutrition/paleo-diet-meal-plan-and-menu

Hoffman, S.J., & Tan, C. (2013). Following celebrities' medical advice: Meta-narrative analysis. *BMJ*, *347*, f7151.

Institute for Integrative Nutrition (IIN) (nd) About Us. Retrieved from www.integrativenutrition.com/about-us

Johnston, J., & Goodman, M.K. (2015). Spectacular foodscapes: Food celebrities and the politics of lifestyle mediation in an age of inequality. *Food, Culture & Society*, *18*(2), 205–222.

Johnston, J., Rodney, A., & Chong, P. (2014). Making change in the kitchen? A study of celebrity cookbooks, culinary personas, and inequality. *Poetics*, *47*, 1–22.

Khamis, S., Ang, L., & Welling, R. (2017). Self-branding,'micro-celebrity' and the rise of Social Media Influencers. *Celebrity Studies*, *8*(2), 191–208.

Lane, S.R., & Fisher, S.M. (2015). The influence of celebrity chefs on a student population, *British Food Journal*, *117*(2), 614–628.

Lewis, T. (2010). Branding, celebratization and the lifestyle expert, *Cultural Studies*, *24*(4), 580–598.

Lewis, T., & Potter, E. (2010). Introducing ethical consumption. In T. Lewis & E. Potter (Eds.), *Ethical Consumption: A Critical Introduction* (pp. 3–24). London: Routledge.

Lohse, B. (2013). Facebook is an effective strategy to recruit low-income women to on-line nutrition education. *Journal of Nutrition Education and Behavior*, *45*(1), 69–76.

Lyson, H.C. (2014). Social structural location and vocabularies of participation: Fostering a collective identity in urban agriculture activism. *Rural Sociology, 79*(3), 310–335.

Matthes, J., & Kohring, M. (2008). The content analysis of media frames: Toward improving reliability and validity. *Journal of Communication, 58*(2), 258–279.

Myrick, J., & Erlichman, S. (forthcoming, 2019). How audience involvement and social norms foster vulnerability to celebrity-based dietary misinformation. *Psychology of Popular Media Culture.*

Nicholas, T. (2014), *The death of expertise.* Retrieved from http://thefederalist.com/2014/01/17/the-death-of-expertise

Nielsen. (2017). *Top six rankings remained consistent in June 2017 digital ratings.* Retrieved from www.nielsen.com/au/en/press-room/2017/top-six-rankings-remained-consistent-in-june-2017-digital-ratings.html

Quinn, L., & McNab, H. (2015, March 16). *Publisher pulls the plug on controversial chef Pete Evans' paleo book for kids …* Retrieved from www.dailymail.co.uk/news/article-2996531/Pete-Evans-Paleo-book-kids-dumped.html

Ramachandran, D., Kite, J., Vassallo, A.J., Chau, J.Y., Partridge, S., Freeman, B., & Gill, T. (2018). Food trends and popular nutrition advice online–implications for public health. *Online Journal of Public Health Informatics, 10*(2). Retrieved from https://journals.uic.edu/ojs/index.php/ojphi/article/view/9306

Robbins, M. (2013). *Love bombing: Oliver James, Susan Greenfield and the rise of data-free celebrity science.* Retrieved from www.theguardian.com/science/the-lay-scientist/2013/jun/04/psychology-guardian-hay-festival

Rose, N.S. (1996). *Inventing our Selves: Psychology, Power, and Personhood.* Cambridge: Cambridge University Press.

Rousseau, S. (2012a). *Food Media: Celebrity Chefs and the Politics of Everyday Interference.* London: Berg.

Rousseau, S. (2012b). *Food and Social Media: You Are What You Tweet.* Lanham, MD: Rowman.

Rousseau, S. (2015). The celebrity quick-fix: When good food meets bad science. *Food, Culture & Society, 18*(2), 265–287.

Salkin, A. (2013). *From Scratch: Inside the Food Network.* New York: Penguin.

Siniscalchi, V., & Counihan, C. (2014). Ethnography of food activism. In C. Counihan & V. Siniscalchi (Eds.), *Food Activism: Agency, Democracy and Economy* (pp. 3–14). London: Bloomsbury.

Styles, A. (2014). *Pete Evans' moveable feast nominated for a daytime emmy.* Retrieved from www.smh.com.au/entertainment/tv-and-radio/pete-evans-a-moveable-feast-nominated-for-daytime-emmy-20140502-37n3h.html

Sydney Morning Herald, The. (2014). 2014 Logies: Full list of winners. Retrieved from www.smh.com.au/entertainment/tv-and-radio/2014-logies-full-list-of-winners-20140428-37ci6.html

Taylor, L. (2015). *Celebrity chef and paleo diet advocate Pete Evans says he's 'thrilled' to release his paleo cookbook for kids independently as an e-book and in print.* Retrieved from www.sbs.com.au/news/pete-evans-thrilled-to-release-kids-paleo-cookbook-independently-as-e-book-in-print

Tran, C. (2015). *Paleo is not a fad… eating like a modern-day tribesman has changed my life.* Retrieved from www.dailymail.co.uk/news/article-2921761/Paleo-not-fad-eating-like-modern-day-tribesman-changed-life-Celebrity-chef-Pete-Evans-defends-food-revolution-sparked-food-war.html

Turner, P.G., & Lefevre, C.E. (2017). Instagram use is linked to increased symptoms of orthorexia nervosa. *Eating and Weight Disorders-Studies on Anorexia, Bulimia and Obesity, 22*(2), 277–284.

Vaterlaus, J.M., Patten, E.V., Roche, C., & Young, J.A. (2015). #Gettinghealthy: The perceived influence of social media on young adult health behaviours. *Computers in Human Behaviour, 45*(1), 151–157.

Willis, O. (2018). *The magic pill: How do the health claims in Pete Evans' paleo diet doco stack up?* Retrieved from www.abc.net.au/news/health/2018–06–21/pete-evans-diet-doco-magic-pill-health-claims-evaluated/9891470

9

'CRAZY FOR CARCASS'

Sarah Wilson, foodie-waste femininity and digital whiteness

Maud Perrier and Elaine Swan

Introduction

Although digital food studies and feminist research on food and femininity proliferate, digital food femininity has yet to be discussed in relation to race, whiteness and, specifically household food waste. Accordingly, in this chapter, we examine the homepage and blog of Sarah Wilson, and an in-post hyperlinked *Mail Online* article, to bring under-researched themes such as digital representations of race, class and food waste to the fore. Such analyses matter because as critical race scholars stress, traditional and digital media representations of race, and whiteness sustain symbolic and material inequalities (Flowers & Swan, 2017; Nakamura, 2008). And as Rachel Slocum and Arun Saldanha (2016) insist, race underpins our social and cultural relations with food in multiple and often unacknowledged ways.

Sarah Wilson is an ex-journalist, ex-editor of Australian *Cosmopolitan* magazine, the first presenter of *Australian Masterchef*, author of ten books, one-time owner of a sugar elimination diet business, 'I Quit Sugar', food-waste activist, and wellness celebrity (Keating, 2014). Wilson has published nine books under her brand 'I Quit Sugar' and one self-help book on anxiety. Her webpage, 'Sarah Wilson', promotes her books, business, brand and lifestyle. It includes a blog, and is connected to platforms such as Facebook, Instagram and Twitter. At the time of writing, Wilson's Facebook page had over 184,000 followers, her Twitter account almost 47,000 followers, and Instagram 231,000 followers. In 2018, she folded her diet business and upped the ante on her association with household food waste reduction in her new book, *I Quit Sugar: Simplicious Flow: The NEW Zero-Waste Eating for an Elegant Life*. Although not as powerful as some food celebrities, Wilson wields some influence on defining eating practices in Australia and globally (Bhagat, 2017; Reynolds, 2015).

In this chapter, first we explore the ways in which Wilson's anti-food waste blog posts reproduce ideals of classed and racialised food femininities. In particular, we focus on verbal and visual texts which represent white femininity and the aestheticisation of food waste. In line with multimodal social semiotics, we use the term 'visual text' to refer to graphics, typeface, digital photographs etc., and 'verbal text', rather than 'written text', for the use of language, linguistics and words (Flowers & Swan, 2018). Second, we examine a *Mail Online* article describing Wilson as 'Crazy for Carcass', a reference to her food rescue tactics of reusing chicken and fish bones from 'plate waste' at dinner parties and restaurants, which was hyperlinked 'in-post' by Wilson on her blog (House, 2015).

Mail Online is the digital version of the *Daily Mail*, a British tabloid, aimed at the lower middle classes, historically with the highest volume of women readers (McRobbie, 1999). Infamous for its sensationalist style, and its sexist, homophobic, racist and Islamophobic reporting, the *Mail* has feminised reporting on fashion, health, healthy food and diets in a section called *Femail*, and promulgates an anti-feminist stance (McRobbie, 1999). The online article we discuss was written by a journalist for the Australian version of *Mail Online* but published by the UK version in October 2015 on the *Femail* menu. *Mail Online* in its UK and Australian versions has millions of readers globally and a history of contradictory coverage of Wilson, both promoting and critiquing her. Indeed, it has a track record of attacking what it sees as diet fads, and yet advocating for others, regulating a normative femininity through critiques of women wellbeing celebrities.

We chose to analyse this hyperlinked article because it represents different versions of Wilson's femininity through visual and verbal texts of domestic food waste, and it provoked direct responses from her. Wilson was exercised enough about the *Mail*'s reporting to respond in two successive, extended blog posts. As hyperlink theorists note, in-post links signal a blogger's reading activity and active conversations with other bloggers, more than blogroll links (De Maeyer, 2013). Although some digital media studies scholars are interested in studies of hyperlinking, interactivity and size of publics (see, for example, De Maeyer, 2013), we are interested in the class between Wilson's representations and those of the hyperlinked article. Moreover, whilst scholars study sharing in food digital cultures (see, for instance, Lupton 2018, 2019), very few have explored contradiction and conflict through hyperlinking. But hyperlinking can disorientate an author's intent, bringing in new screens, content, and language, potentially destablising the blog's composition, coherence and authorial voice (Djonov, 2007).

Curated for self-expression, and promotion, a blog is a means for bloggers to construct themselves as 'taste leaders' (McQuarrie, Miller, & Phillips, 2013). Wilson's blog promotes her as a 'taste leader' in sugar-free consumption and domestic food waste avoidance, and advertises her products, classed aesthetics and embrace of a low consumption lifestyle associated with the so-called 'voluntary simplicity movement' (Aguilar, 2015). We examine how Wilson's femininity changed visually and politically through the remediation of food waste outside of her digital community. In particular, we explore how what we call Wilson's 'foodie waste' aesthetic and persona mobilises her middle-class femininity through a visual stylistics of discipline, cleanliness and elegance which symbolically reinforces her whiteness through cultural and historic associations of hygiene and control (Anderson, 2000; Berthold, 2010; Roth-Gordon, 2011; Swan, 2012). Moreover, her blog evacuates the enduring connection of food waste and leftovers with dirt, impurity, stigma, race and class (Beizer, 2017; Romero, 1994; Whitthall, 2017).

Writing about dumpster diving – rescuing discarded but edible food waste from skips and bins – Alex Barnard (2016) notes that waste creates strong pejorative emotional and moral meanings in Western culture. Turning to related discussions on hygiene, race and whiteness, Dana Berthold (2010) shows the racist heritage of 'purity ideals' in contemporary hygiene discourse and excavates the links between cleanliness, civility, class, whiteness and moral virtue. In relation to raced and class histories of leftovers, historians reveal how slaves in the US and racialised domestic servants in America and Australia were expected to eat leftovers, reinforcing racial and class hierarchies, with evidence that in South America this practice continues with leftovers 'gifts' in lieu of pay rises or requests for holidays (Covey & Eisnach, 2009; Jones, 2004; Romero, 1994; Walden, 1995). Discussing the pejoration of doggy bag use for recuperating plate waste in contemporary France, Janet Beizer (2017) notes that doggy bags are still stigmatised because of the association of leftovers with free meals for the urban poor in late nineteenth-century Paris. Also discussing the racialised stigmatisation of rescuing discarded food, Barnard (2016) argues racially minoritised groups undertake their practices at night to avoid social shame.

As we discussed, digital food studies as a field has yet to grapple in any meaningful way with how digital representations of food are irrevocably entangled with race, and with white femininity. In this chapter, we work from the premise that the classed and racialised cultural economy of hygiene, cleanliness, purity and food waste inflect relations to household leftovers, with their enduring racialised representation as dirty and distasteful and the stigmatisation of people who are forced to eat leftovers. Hence, we map how visual and verbal representations of food waste – digitally mediated food waste – are contingently articulated to classed and racialised stigma surrounding eating leftovers and the extent to which it is remediated in Wilson's foodie waste persona and aesthetics.

Mediated food femininities

Digital food cultures scholars stress the significance of digital and new media, and in particular the food blogosphere as a significant site of feminised media production and gender construction including digital femininity and food femininities (Contois, 2017; Dejamanee, 2016; Lupton, 2018, 2019; Rodney, Cappeliez,

Oleschuk, & Johnson, 2017). The first to coin the term 'food femininities', Kate Cairns and Josée Johnston (2016), highlight how femininities are performed in multiple, shifting and hierarchised ways. Their main finding is the narrow range of 'acceptable' food femininities within which women are incited to care about food, but not too much. Other media feminists scope the emergence of a different kind of femininity, the 'new' housewife (Jensen, 2013; Nathanson, 2013). They find renewed interest in domestic crafts such as sewing, knitting, preserving and growing food and understood in Britain as belonging to the post-war grandmothers' generation. Such a revival constitutes a re-traditionalisation of domesticity, reliant on the nostalgisation of a bygone age where women's identities were tied up with the home (Hollows, 2008; Nathanson, 2013). Scholars note the idealisation of the frugal housewife as a distinct iteration of traditional femininity within the specific sociohistorical circumstances of austerity (Jensen, 2013). This work problematises the figure of the 'thrifty housewife', through which being thrifty is positioned as a marker of middle-class respectability, concealing privilege and central to the moral economies of the self in late capitalism (Jensen, 2013).

Research focused more specifically on digitally mediated food femininities has largely studied food blogs and the feminist politics of food porn, healthy, clean eating and the balance between care, discipline and pleasure (Contois, 2017; Dejamanee, 2016; Lupton, 2018, 2019). Building on the concept of food femininities, Alexandra Rodney et al. (2017) mobilise the term 'domestic goddess' to denote a popular digital representation based on domestic responsibility, fulfilment and corporeal discipline. Their research foregrounds femininity and class, rather than race and whiteness. In contrast, Emily Contois' (2017), writing about healthy food blogs, shows how digital white femininity is reproduced through the highly curated aesthetics of represented food, the white hyper-feminine bloggers' bodies and their affluent heteronormative lifestyles.

Household food waste studies

Digital food femininity has yet to be discussed in relation to household food waste. Indeed, household food waste studies tend to research sociomaterial household practices rather than cultural and digital constructions of food waste. Moreover, they give little attention to how gender, race and class shape understandings and practices of food waste (Evans, Campbell & Murcott, 2013; see Cappellini & Parsons, 2012; Watson & Meah, 2012 for gendered analyses, although these too neglect race). This scholarship highlights that waste is not a fixed, clear-cut and self-evident category but ontologically unstable. Food leftovers – a prominent element in household waste reduction strategies – are particularly ambiguous in nature, 'neither meal nor ingredients, neither fresh nor completely spoiled; as such, they do not belong on the plate but neither do they yet belong in the waste bin' (Cappellini & Parsons, 2012, 123). In response to the neglect of gender in food waste studies, Carly Fraser and Kate Parizeau (2018) call for feminist analyses of what they call 'household food waste work'. However, this work also tends to neglect consideration of how gender, race and class hierarchies inflect the cultural meanings attached to food waste, and of the imagined and lived proximities of minoritised groups to food waste and cultural associations of dirt and lack of hygiene with race and class.

To date, there are few studies of digitally mediated food waste. Most digital food waste studies examine policy initiatives. The few studies of digital household food waste stay largely at the level of analysing language (see, for example, Mirosa, Yip, & Lentz, 2018) rather than other elements such as visual texts, interactivity and infrastructure (see Lupton, 2018, 2019; Nakamura, 2008). Thus far, Finnish academics Elina Närvänen, Nina Mesiranta, Ulla-Maija Sutinen and Malla Mattila (2018) provide the most sustained attention to digital representations of household food waste avoidance. Their main finding is that whilst food has positive and pleasurable meanings, food waste connotes the negative and unwanted. Therefore, campaigners aestheticise food waste and present leftovers in imaginative and visually appealing ways.

Although this literature provides new understandings of femininities and food, scholars give insufficient attention to the racialisation of food femininities. The scholarship on food femininities ignores how contemporary constructions of domesticity assume whiteness and privileges an analytic frame of postfeminism, characterised by representations of the "'pastness" of feminism and the neoliberal "focus upon individualism, choice and empowerment"' (Gill, 2007 cited in Dejamanee, 2016, p. 432).

In this chapter, we contribute to the scholarship on food femininities and digital food studies by exploring how Wilson's food waste persona differs from these new domestic femininities – the thrifty housewife and domestic goddess. Moving away from the postfeminist frame allows us to attend to the cultural connotations of food waste with dirt, hygiene, race and class and the gendered, racialised and classed aestheticisation of food waste. We argue these categories significantly shape the authority and identities adopted by those who attempt to disrupt what counts as food waste, as our analysis of Wilson's blog and its reverberations demonstrates.

Methodology

Our methodological approach to studying Sarah Wilson's social media drew on studies of digital whiteness, particularly in relation to food (Flowers & Swan, 2017; Swan, 2017) and feminist multimodal social semiotics (Adami, 2014; Swan, 2017). Broadly speaking, the latter is an approach in which different elements of meaning-making, such as visual and verbal elements, are analysed for their distinctive ways of representing the world to interrogate ideologies about gender, class and race (Swan, 2017). Accordingly, in our first stage, we surveyed Wilson's social media, noting images of food waste and femininity, and taking screenshots of key food waste blog posts, her website's home page and the hyperlinked *Online*

Mail article discussing Wilson. In our more detailed analysis, we analysed a sample of the visual and verbal texts coding whiteness, femininity, class, elegance, cleanliness, race, and food waste. In analysing race, we drew on Richard Dyer's (2004) visual whiteness framing and Stuart Hall's argument that race is signified through various visual tropes including coloured objects, 'foreign' scripts, and animals (Hall, 1997 cited in Flowers & Swan, 2017). Regarding food waste, we noted leftovers, food being thrown in bins, dishes made from leftovers, spoiled and mouldy food waste, discarded edible raw and cooked foods (inspired by Thompson & Haigh, 2017).

Sarah Wilson's digital white femininity

This section provides a brief description of Wilson's homepage, followed by a discussion of how this constructs digital white femininity. A viewer visiting the homepage of Wilson's website encounters a dynamic banner depicting five large, professionally taken, carefully crafted digital photographs of Wilson: slim, smiling, tanned, face flawlessly well-made up, hair long and streaked blonde, looking youthful, toned, athletic and active (Wilson, 2019). Four of the five images show her out of doors: three by the sea and one on the beach, as is common for Australian fashion shots. Wilson is depicted typically as wearing short dresses, or a favourite pair of green shorts which get their own blogpost, her tanned athletic but feminine thighs on display. Wilson's online persona as a woman who does not have a home or cook for her family provides us with an atypical entry point into digital food femininity discourses.

The images of Wilson on the homepage reproduce an idealised white femininity found across films, advertising and food digital media but one intercut by Australian histories of race and femininity (Bhagat, 2017). Critical whiteness scholars argue that lighting technologies brighten white skin and blonde hair, and together with flawless make-up create milky white skin and an ethereal glow key to white middle-class femininity (Dyer, 2004; Swan, 2012). Wilson's images draw on these tropes for the most part (her skin is lightly tanned rather than milky white). But this does not undermine her white femininity. As Australian feminists highlight, there are specificities to the cultural production of white Australian femininity in the light of British colonialism and associated racialisation processes (Ahmed, 1998; Hamilton, 2006; Maynard, 1999; Nash, 2018). Since early colonisation, Australian white femininity has been associated in the media with tanned, toned, outdoor athleticism: a 'new kind of womanhood', robust, self-reliant, and health-loving 'aligned with nationalist and eugenicist ideas of the interwar period (Maynard, 1999, 202). As Meredith Nash writes, 'today, the blonde haired, blue-eyed, tanned bikini girl is one of the most salient visual representations of white Australian beach culture, idealised femininity and national identity' (2018, p. 592). This idealisation of white Australian femininities relies on the erasure of Aboriginal dispossession and racial segregation (ibid.). As with the images of a tanned Wilson, the perfect Australian tanned 'beach body' allows white women to acquire 'colour' while still maintaining white privilege (Ahmed, 1998). Sara Ahmed notes that:

> the perfectability of tanned skin remains bound up with notions of hygiene. Tanned colour is clean colour and is hence immediately distinguished from the infectability of being-Black. Here, the techniques for tanning the body provide the means for cleanliness ('hygienic perfection'): the tanned surface-skin denotes the moral hygiene of the white soul (underneath).
>
> *(1998, p. 58)*

Hence, we can understand the photographic images of Wilson on her blog as promoting not only a classed, cultivated, disciplined body, but also a bodily and moral hygiene. Indeed, these photographs are aesthetically isomorphic with the visual texts on the homepage – sparse verbal text, limited typefaces, cohesive colour palette and well-spaced out text. Her aesthetics align with middle-class minimalism and sophisticated simplicity, connoting discipline, cleanliness and control, and resonating with other digital 'body/food' media (Lupton, 2019), together with a classed elegance, distancing Wilson's femininity from the 'dross of the domestic' (Swan, 2012, 198). Indeed, minimalism features strongly in Wilson's online persona as reflected by verbal texts on the homepage which describe her as 'author, minimalist and entrepreneur'. Wilson's brand is 'simplicious': or as she puts it in one blogpost in 2011, 'It's simple and elegant. Waste Less.'

Simplicity aesthetic?

In the next section, we turn to Wilson's food waste blog posts to show how she aestheticises eating leftovers into a practice of discipline that bolster the white middle-class construction of femininity as controlled and clean. Users find her food waste blog posts via an interactive search button on a small menu on the right of the homepage. Food waste is not a category on the menu but blog posts about food waste are hashtagged #foodwaste, #whatIeat and #simplicious. Wilson uses these hashtags across her Instagram, Facebook and Twitter accounts, enabling her to establish and join like-minded communities of users on social media. As is typical of blogs as a genre, Wilson's posts are word centred, although visual images of dishes, often reposted and repurposed from her books, pepper the verbal texts. Overall, at the time of our research, 90 of Wilson's blog posts had been tagged #foodwaste and these posts had been shared between 50 and 100 times by her viewers on other social media platforms such as Twitter and Facebook. Although this is very low sharing compared to followers of very well-known food and wellbeing celebrities and influencers, it mirrors other Australian food-related celebrity social media such as that of paleo chef and television presenter Peter Evans (see Chapter 8).

Unlike male celebrity food waste activists such as the UK's Hugh Fearnley-Whittingstall, who mobilises TV and social media to advocate for waste reductions across the whole food system, drawing on a masculine persona of authority and expertise, Wilson's campaign is clearly feminised. She constantly references her everyday life, staying at the level of household waste and feminised food work, and drawing on feminised genres such as recipes, confessionals and tips (Craig, 2018; Swan, 2017). Although like other food waste bloggers, Wilson's content invokes a range of constructions of food waste, from thrift, sustainability, ethics, grandparent traditions and responsibility to others facing hunger (Närvänen et al., 2018), her project promotes an individualistic approach to household food waste. Her main focus is repurposing edible food waste through micro-food rescue practices, including re-creating new meals from plate waste, and cooking the edible parts of foods that are usually discarded. Ignoring the ways in which minoritised groups can be economically forced into eating leftovers, the blog constructs preparing and eating food waste visually and verbally as fun, fashionable, and even 'sexy'.

In line with Wilson's homepage, the digital images of the leftover meals reproduce a middle-class aesthetics of wholesome elegance, simplicity, cleanliness and discipline. Thus, the food waste posts do not depict images of food waste, meal preparation, kitchen equipment or even people eating food. Rather, images draw on middle-class white foodie visual tropes of simplicity: dishes of vibrantly coloured, mismatched tableware and wholesome food placed on a natural-looking surface (hessian, wood, cotton) and photographed from above, communicating casually humble home-made food, upcycling and a stylised spontaneity, and authenticity (Taylor & Keating, 2018). This is a constructed fauxcasual aesthetic, invoking a wider set of active themes of nostalgia, frugality and self-imposed restraint found in food popular culture, and downshifting lifestyle media (Wincott, 2016). Wilson's representation conceals the classed resources deployed to achieve this domestic aesthetic and the material labour involved by making the dish appear effortlessly made (Hollows, 2003) and reproduces frugality as a lifestyle, rather than the lived realities of daily essential 'belt-tightening' (Wincott, 2016). Hence, she produces 'differentiated' food waste strategies, distanced from material need and highlighting their aesthetics.

Producing digital foodie-waste femininity

This visual imagery sidesteps the association of food waste with stigma, poverty and dirt, reproducing what we call 'foodie waste' – a white bourgeois feminised foodie aesthetic and ethical practice premised on symbolic and material privilege. As identified above, Wilson's food waste posts reproduce a sense of cleanliness and control, and a wholesome, sustainable aesthetic associated with white middle-class femininity (Logan, 2017). Ideas about cleanliness and dirt have been used to differentiate women by class, gender and race, symbolically and materially (Anderson, 2000; Swan, 2012). White women have and continue to be protected from the 'defiling contact with the sordid or disordered parts of life' (Davidoff, 1973 cited in Swan, 2012, p. 187). Furthermore, colonialism entwined notions of cleanliness and being civilised (McClintock, 1995) with the inculcation of 'disciplines and routines of personal hygiene' part of the 'civilising' project of British imperialism, associated with self-refinement (Shove, 2003) and which in Australia was a way in which Indigenous people and Asians were racially excluded (Bashford, 1998). As Jennifer Roth-Gordon writes:

> understandings of race are centrally preoccupied with the ability to discipline ourselves and exhibit proper bodily control. White bodies are defined by their 'natural' proclivity towards discipline and their capacity for refinement and control, in contrast to non-white bodies that can be defined by … a lack of personal hygiene or bodily cleanliness.
>
> *(2011, p. 213)*

The notion of cleanliness and discipline is further amplified as an aesthetic and ethical practice through Wilson's 'simplicious flow' philosophy that constructs eating leftovers into a quest for purity, simplicity, self-imposed discipline and morality. Indeed, these themes interrelate across all her interests in simplicity, control and purity from anti-sugar, pro hiking, clean living, 'healthy' cooking techniques and ethical consumption.

Highlighting discipline and morality, Wilson presents wasting food as lazy and hypocritical. Her verbal texts express a commitment to using 'every last bit of food'. She claims to enjoy using the very last 'dredges' of food in her fridge to make a packed lunch before she goes on holiday. Conspicuously absent from her blog are the times where she and her readers could not be bothered or did not have time to undertake the labour-intensive food waste work. The foodie waste practices that Wilson encourages echo self-disciplining techniques visible in make-over shows that offer participants the promise of attaining a respectable bourgeois feminine subjectivity if they eat, look and clean better (McRobbie, 2004; Nathanson, 2013).

Minimalist aesthetics and voluntary simplicity practices are cultural and moral practices reserved for the elite, the latter resulting from already possessing too much (Logan, 2017). The achievement of a minimalist home design suits either the childfree or those who can rely on the labour of either middle-class housewives, domestic racially minoritised or white working-class servants (Leslie & Reimer, 2003). Similar to voluntary simplicity practices of decluttering which reject abundance in favour of asceticism, foodie waste minimalism needs to be understood as part of resourcing ethical middle-class white selves.

Wilson's minimalist food waste persona draws on the historical construction of white middle-class femininity as requiring control and discipline. Through its conversion into an aesthetic of moral purity, food waste loses its stigma. Indeed, our reading of her blog suggests that accruing value to what is considered waste relies on occupying a particular classed and racialised status: 'The capacity to convert the discarded into something of value depends on the authority of those who do the re-evaluation' (Negrin, 2015, p. 200). Wilson's foodie waste minimalism is characterised by a disciplined and moral practice that is distinctively recognisable as middle-class white femininity to her readers.

'Crazy for carcass': unstable foodie-waste femininity

In this section, we show the limits to the legibility of Wilson's attempt to reclaim eating food waste as an elegant foodie practice beyond her blog. Thus, we argue that hyperlinked *Mail Online* article disrupts Wilson's white foodie-waste femininity. The placement of the hyperlink so prominently on her blog post on October 2015, and referenced again a week later, shows that she wishes to foreground the article. Our reading of her response suggests that she is not simply doing this to promote her brand but is aggravated by some content. For instance, she describes being 'taken down' and references being 'trolled'.

To go into more detail: when a viewer clicks from the highlighted verbal hyperlink to the new *Mail Online* page, they see a headline in huge black typeface:

'People see rubbish, I see opportunity': Sarah Wilson reveals she takes leftovers from strangers' plates in restaurants – and she wants YOU to eat scraps too.

Underneath the headline, the *Mail Online* reproduces quotes from Wilson, foodie images from her book, a video of Wilson, together with the its own commentary – both positive and provocative – and unappealing stock images of food waste from Getty. Whilst the article promotes Wilson, our reading suggests that images and verbal text destabilise her foodie-waste femininity, poking fun at some of her practices such as repurposing bones from strangers' and friends' plates, reusing tea-bags and asking for doggie bags. Indeed, in the captions below the images, it is noticeable that the terms 'germs', 'barely touched duck leg', 'ugly veg' and 'fish carcass bag' are all in scare quotes, highlighting the contested status of this repurposed food and raising suspicion of Wilson's claim they are 'perfectly safe' to consume.

But the most arresting provocation comes in the form of three stock photographic images and the subheading describing Wilson as 'crazy for carcass'; the *Mail* using a feminised pejorative implying Wilson is unhinged and out of control (Montell, 2018). One image shows a white woman's hand throwing an egg into a bin heaving with spoiled food. Whilst it can be argued that it is an aestheticised, clean image of food waste, Wilson does not include images of spoiled food waste or bins on her blog. The other two images depict a half-eaten chicken carcass and a chicken bone. The unstylised images are shot in harsh bright light against white decontextualised backgrounds, the bones and meat in lacklustre colours, appearing very unappetising, the opposite of visually appealing foodie photographs of designed and choreographed dishes of ripe, plump and fresh food (Taylor & Keating, 2018). Culturally, farm chickens are feminised, seen as 'unnatural, tame and confined', in opposition to masculinised, wild, free animals, and deemed to have 'lives [which] appear too slavishly, too boringly, too stupidly female' (Davis, 1995, 196). Moreover, they represent the high-tech industrialised throwaway food, their carcass meat seen as 'bad' meat on TV programmes such as Jamie Oliver (Phillipov, 2017).

Moreover, the hyperlinked article brings in a different community of users. In the interactive comments box, 13 *Mail Online* viewers make abrupt, denigratory comments about Wilson's lack of hygiene. This number of comments is hardly an excessive reaction in terms of volume, but most express intense disgust, describing Wilson as 'gross', a term the polar opposite of clean, suggesting feelings of repulsion (Connor, 2011). The *Mail Online* viewers represent a counter to the typical 20 to 50 grateful commentators on Wilson's blog, many of whom thank her for her insights, share their own tips and agree with the practice of repurposing bones.

The hyperlinked article challenges Wilson's aestheticisation of eating leftovers and her white middle-class femininity. First, the images of the carcass, chicken bone and the bin bring a reality to food waste missing from Wilson's highly stylised, sensory 'foodie-waste' images and the pared-down elegance of her homepage and blog. Second, the combination of the images of the chicken bones, critical reader comments and the *Mail's* verbal ironisation congeal to suggest that Wilson's actions are dirty and uncivilised because of classed and racialised associations of untouchability and pollution. As Edward Witthall notes 'food half-consumed, touched by teeth, hands and saliva, comes uncomfortably close to human waste' (2016, p. 105). The stigma surrounding eating leftovers is enhanced by their contaminated status: 'the idea of eating food that is potentially polluted by the mouth or touch of the Other, or tainted because of spoilage or unhygienic practices, is closely bound to the concept of the leftover' (Beizer, 2017, p. 379). As Helen Veit states more succinctly, leftovers 'have always been uncomfortably close to garbage' (2015). And waste is frequently associated with morally suspect groups. Hence, disgust slides from the spoiled food and chicken meat to Wilson, and destabilises her claims to moral superiority, purity and cleanliness. But more than this, its verbal and visual texts racialise and class her through the associations of dirt, disease, lack of hygiene, contamination and stinginess. In so doing, the *Mail Online's* article challenges her attempt to construct eating leftovers as good, disciplined and elegant, and her claims to frugality and purity.

The *Mail's* visual representation of food waste reveal the instability of Wilson's aestheticisation of leftovers. The *Mail* re-semioticise leftovers as dirty and unhygienic, and in so doing, reveal the instability of her authoritative claims to define morally and materially what constitutes edible food waste. Through this process and the racialised and classed connotations of a lack of body control, decorum and hygiene, Wilson temporarily 'loses' her middle-class symbolic white femininity (Roth-Gordon, 2011). In positioning Wilson as 'crazy', her authority to be a taste leader and proclaim what is edible, clean and ordered is challenged.

Conclusion

In this chapter, we explore how Sarah Wilson's food waste posts constructs eating leftovers as a bourgeois foodie practice and how her attempts to aestheticise eating food waste are challenged outside of her immediate blog ecology. We identified that despite her food waste blog articulating a distinctively and reassuringly white middle-class foodie-waste femininity, her eat food waste campaign is constructed as threatening and ridiculous by the *Mail* and some of its readers. Our analysis emphasises how her foodie-waste femininity mobilises repertoires of white middle-class control, discipline and purity rather than the pleasure and care that the postfeminist frame deployed by previous scholars of food and femininity highlights. We argued that her privilege enables her to question what counts as waste and she uses these resources to create a foodie waste persona and aesthetic. While the unsavoury associations of leftovers are concealed from her website, the spectre of leftovers as 'dirty food' resurfaces in the *Mail*'s portrayal of Wilson as 'too extreme'. But she attempts to reclaim moral superiority on her own blog by directly responding to the article.

Feminists underline how the control of women's appetites and bodies characterise mediated food femininities and our analysis shows this extends to not wasting food. Wilson's 'simplicious minimalism' is a performance of control shaped by classed and racialised ideals of cleanliness, discipline and purity. Our chapter shows the ways in which Wilson's foodie waste persona and the scorn it generates is premised on the racialised and classed histories of hygiene and cleanliness and of leftovers as contagious and contaminated. Furthermore, we extend digital food studies by showing how studying the hyperlinked article enable us to map the instabilities in Wilson's foodie-waste femininity and in so doing underlining how hyperlinked texts can disrupt coherence in ways that go beyond studies of verbal and visual texts.

While we focus on an Australian case study, national and international anti-food waste campaigns are increasingly being led by celebrities (Craig, 2018). Yet few digital food scholars comment on how gendered, classed and racialised meanings are embedded in the way household food waste and eating leftovers are mediated. Furthermore, the way Sarah Wilson's foodie waste persona has generated interest beyond Australia and her usual readership suggests that repurposing food waste is increasingly central to digital food cultures, with the classed and racialised aestheticisation of food waste playing a central role in mainstreaming its appeal and generating captive audiences. In so doing, we show how digital food culture continues to associate whiteness and class with notions of purity, discipline, cleanliness and control in ways which reproduce symbolic and material inequalities.

Further reading

Finnish academics' website on food waste research: http://wastebustersfinland.blogspot.com/

The authors' website on critical race and feminist studies of food work and pedagogies: www.servingsblog.com

Dr Lisa Nakamura's website on digital studies, race and gender: https://lisanakamura.net/

References

Adami, E. (2014). Social taste and meaning at glance: A multimodal framework of aesthetics in webtexts. National Centre for Research Methods Working Paper. London: Institute of Education.

Aguilar, J. (2015). Food choices and voluntary simplicity in intentional communities: What's race and class got to do with it? *Utopian Studies, 26*(1), 79–100.

Ahmed, S. (1998). Animated borders: Skin, colour and tanning. In J. Price and M. Shildrick (Eds.), *Vital Signs: Feminist Reconfigurations of the Biological Body* (pp. 45–65). Edinburgh: Edinburgh University Press.

Anderson, B. (2000). *Doing the Dirty Work? The Global Politics of Domestic Labour.* New York: Zed Books.

Barnard, A.V. (2016). Making the city 'second nature': Freegan 'Dumpster divers' and the materiality of morality. *American Journal of Sociology, 121*(4), 1017–1050.

Bashford, A. (1998). Quarantine and the imagining of the Australian nation. *Health, 2*(4), 387–402.

Beizer, J. (2017). Why the French hate doggie bags. *Contemporary French Civilization, 42*(3–4), 373–389.

Berthold, D. (2010). Tidy whiteness: A genealogy of race, purity, and hygiene. *Ethics & the Environment, 15*(1), 1–26.

Bhagat, R. (2017, July–December). Spare rib in *Kill Your Darlings.* Retrieved from https://Killyourdarlings.Com.Au/Category/Commentary

Cairns, K., & Johnston, J. (2015). *Food and Femininity.* London: Bloomsbury.

Cappellini, B., & Parsons, E. (2012). Practising thrift at dinnertime: Mealtime leftovers, sacrifice and family membership. *The Sociological Review, 60,* 121–134.

Connor, S. (2011). *Smear campaigns.* Retrieved from http://stevenconnor.com/smearcampaigns.html

Contois, E. (2017) Healthy food blogs: Creating new nutrition knowledge at the crossroads of science, foodie lifestyle, and gender identities. In B. Bock and J. Duncan (Eds.), *Gendered Food Practices from Seed to Waste: Year Book of Women's History* (pp. 129–143). Amsterdam: Hilversum Verloren.

Covey, H.C., and Eisnach, D. (2009). *What the Slaves Ate: Recollections of African American Foods and Foodways From the Slave Narratives.* Santa Barbara, CA: Greenwood.

Craig, G. (2018). Sustainable everyday life and celebrity environmental advocacy in *Hugh's War on Waste. Environmental Communication, 13*(6), 775–789.

Davidoff, L. (1973) Domestic service and the working-class life-cycle. *Bulletin of the Society for the Study of Labour History, 26,* 10–12.

Davis, K. (1995). Thinking like a chicken: Farm animals and the feminine connection. In L. Birke (Ed.) *Animals and Women: Feminist Theoretical Explorations* (pp. 192–221). Durham: Duke University Press.

Dejamanee, T. (2016). 'Food porn' as postfeminist play: Digital femininity and the female body on food blogs. *Television & New Media, 17*(5), 429–448.

De Maeyer, J. (2013). Towards a hyperlinked society: A critical review of link studies. *New Media & Society, 15*(5), 737–751.

Djonov, E. (2007). Website hierarchy and the interaction between content organization, webpage and navigation design: A systemic functional hypermedia discourse analysis perspective. *Information Design Journal, 15*(2), 144–162.

Dyer, R. (2004). *Heavenly Bodies: Film Stars and Society* (2nd ed.). London, Routledge.

Evans, D., Campbell, H., & Murcott, A. (2013). A brief pre-history of food waste and the social sciences. *The Sociological Review, 60*(S2), 5–26.

Flowers, R., & Swan, E. (2017). Seeing benevolently: Representational politics and digital race formation on ethnic food tour webpages. *Geoforum, 84*, 206–217.

Flowers, R., & Swan, E. (2018). The welcome dinner project. In M. Phillipov & K. Kirkwood (Eds.), *Alternative Food Politics: From the Margins to the Mainstream* (pp. 95–112). London: Routledge.

Fraser, C., & Parizeau, K. (2018). Waste management as foodwork: A feminist food studies approach to household food waste. *Canadian Food Studies, 5*(1), 39–62.

Hamilton, M. (2006). A girl cannot be beautiful unless she is healthy: Nationalism, Australian womanhood, and the pix beach girl quests of World War II. In L. Boucher, J. Carey & K. Elllinghaus (Eds.), *Historicising Whiteness: Transnational Perspectives on the Construction of an Identity* (pp. 234–243). Melbourne: RMIT Publishing.

Hollows, J. (2003). Feeling like a domestic goddess: Postfeminism and cooking. *European Journal of Cultural Studies, 6*(2), 179–202.

Hollows, J. (2008). *Domestic Cultures*. Maidenhead: Open University Press.

House, L. (2015) *'People see rubbish, I see opportunity': Sarah Wilson reveals she takes leftovers from strangers' plates in restaurants – and she wants YOU to eat scraps too*. Retrieved from www.dailymail.co.uk/femail/article-3261249/Sarah-Wilson-s-mission-combat-wastage-sees-taking-leftovers-strangers-plates-restaurants-wants-eat-scraps-too.html

Jensen, T. (2013). Austerity parenting. *Soundings, 55*(55), 61–71.

Jones, J. (2004). Working for the white people. *Balayi: Culture, Law and Colonialism, 6*, 1–8.

Keating, E. (2014) *Sweet success: How Sarah Wilson turned her I Quit Sugar blog into a global phenomenon*. Retrieved from www.smartcompany.com.au/entrepreneurs/influencers-profiles/sweet-success-how-sarah-wilson-turned-her-i-quit-sugar-blog-into-a-global-phenomenon/

Leslie, D., & Reimer, S. (2003). Gender, modern design, and home consumption. *Environment and Planning D: Society and Space, 21*(3), 293–316.

Logan, D. (2017). The lean closet: Asceticism in postindustrial consumer culture. *Journal of the American Academy of Religion, 85*(3), 600–628.

Lupton, D. (2018). Cooking, eating, uploading: Digital food cultures. In K. LeBesco & P. Naccarato (Eds.), *The Handbook of Food and Popular Culture* (pp. 66–79). London: Bloomsbury.

Lupton, D. (2019). Vitalities and visceralities: Alternative body/food politics in new digital media. In M. Phillipov & K. Kirkwood (Eds.), *Alternative Food Politics: From the Margins to the Mainstream* (pp. 151–169). London: Routledge.

Maynard, M. (1999). Living dolls: The fashion model in Australia. *The Journal of Popular Culture, 33*(1), 191–205.

McClintock, A. (1995). *Imperial Leather: Race, Gender and Sexuality in the Colonial Contest*. New York: Routledge.

McQuarrie, E.F., Miller, J., & Phillips, B.J. (2012). The megaphone effect: Taste and audience in fashion blogging. *Journal of Consumer Research, 40*(1), 136–158.

McRobbie, A. (1999). *Feminism v the TV blonde*. Inaugural lecture, Goldsmiths College, University of London.

McRobbie, A. (2004). Notes on 'What not to wear' and postfeminist symbolic violence. *Sociological Review, 52*(2), 99–109.

Mirosa, M., Yip, R., & Lentz, G. (2018). Content analysis of the 'clean your plate campaign' on Sina Weibo. *Journal of Food Products Marketing, 24*(5), 539–562.

Montell, A. (2018, May 1) *A brief yet fascinating history of the word 'crazy'*. Retrieved from https://thethirty.byrdie.co.uk/etymology-crazy-sexist-words

Nakamura, L. (2008). *Digitizing Race: Visual Cultures of the Internet*. Minneapolis: University of Minnesota Press.

Närvänen, E., Mesiranta, N., Sutinen, U.M., & Mattila, M. (2018). Creativity, aesthetics and ethics of food waste in social media campaigns. *Journal of Cleaner Production, 195*, 102–110.

Nash, M. (2018). White pregnant bodies on the Australian beach: A visual discourse analysis of family photographs. *Journal of Gender Studies, 27*(5), 589–606.

Nathanson, E. (2013). *Television and Postfeminist Housekeeping: No Time for Mother*. London: Routledge.

Negrin, L. (2015). The contemporary significance of 'pauperist' style. *Theory, Culture & Society, 32*(7–8), 197–213.

Phillipov, M. (2017) *Media and Food Industries: The New Politics of Food*. London: Palgrave Macmillan.

Reynolds, R.C. (2015) *Quit sugar, go paleo, embrace 'clean food': the power of celebrity nutrition*. Retrieved from https://theconversation.com/quit-sugar-go-paleo-embrace-clean-food-the-power-of-celebrity-nutrition-38822

Rodney, A., Cappeliez, S., Oleschuk, M., & Johnston, J. (2017). The online domestic goddess: An analysis of food blog femininities. *Food, Culture & Society, 20*(4), 685–707.

Romero, M. (1994). Transcending and reproducing race, class and gender hierarchies in the everyday interactions between Chicana private household workers and employers. In V. Demos & M. Texler-Segal (Eds.), *Ethnic Women: A Multiple Status Reality* (pp. 135–144). Dix Hills, NY: General Hall.

Roth-Gordon, J. (2011). Discipline and disorder in the whiteness of mock Spanish. *Journal of Linguistic Anthropology, 21*(2), 211–229.

Shove, E. (2003). *Comfort, Cleanliness and Convenience: The Social Organization of Normality*. Oxford: Berg.

Slocum, R., & Saldanha, A. (2016) (Eds.) *Geographies of Race and Food: Fields, Bodies, Markets*. London, Routledge.

Swan, E. (2012) Cleaning up? Transnational corporate femininity and dirty work in magazine culture. In R. Simpson, N. Slutskaya, P. Lewis, & Höpfl, H. (Eds.), *Dirty Work: Concepts and Identities Reality* (pp. 182–202). Houndsmills: Palgrave Macmillan.

Swan, E. (2017). Postfeminist stylistics, work femininities and coaching: A multimodal study of a website. *Gender, Work & Organization, 24*(3), 274–296.

Taylor, N., & Keating, M. (2018). Contemporary food imagery: Food porn and other visual trends. *Communication Research and Practice*, 1–17.

Thompson, K., & Haigh, L. (2017). Representations of food waste in reality food television: An exploratory analysis of Ramsay's kitchen nightmares. *Sustainability, 9*(7), 1139.

Veit, H. (2015) *An economic history of leftovers*. Retrieved from www.theatlantic.com/business/archive/2015/10/an-economic-history-of-leftovers/409255/

Walden, I. (1995). 'That was slavery days': Aboriginal domestic servants in New South Wales in the twentieth century. *Labour History: A Journal of Labour and Social History, 69*, 196–209.

Watson, M., & Meah, A. (2012). Food, waste and safety: Negotiating conflicting social anxieties into the practices of domestic provisioning. *The Sociological Review, 60*, 102–120.

Whittall, E. (2017). Performing leftovers: On the ecology of performance's remains. *Performance Research, 22*(8), 99–106.

Wilson, S. (2018). *I Quit Sugar: Simplicious Flow: The New Zero-Waste Eating for an Elegant Life*. Sydney: Macmillan Australia.

Wilson, S. (2019). [Home page]. www.sarahwilson.com/

Wincott, A. (2016). The allotment in the restaurant: The paradox of foodie austerity and changing food values. In P. Bennet & J. McDougall (Eds.), *Popular Culture and the Austerity Myth: Hard Times Today* (pp. 44–57). London: Taylor & Francis.

PART 4

Spatialities and politics

10

ARE YOU LOCAL?

Digital inclusion in participatory foodscapes

Alana Mann

Introduction

> Consider the places and spaces where you acquire food, prepare food, talk
> about food, or generally gather some sort of meaning from food. This is
> your foodscape.
>
> *(Mackendrick, 2014, 16)*

Our foodscapes are increasingly digital. Many of us now source and purchase
food online, download grocery lists and recipes, and share our meals through
photo-sharing platforms such as Instagram and Facebook – all of which can
significantly influence our food choices. Lewis and Phillipov (2018) assert that
digital food practices have become so habitual that they often pass 'unnoticed
in people's daily lives', as in the case of on-demand food delivery apps focusing
on delivery speed, cashless transactions, food quality and customer relationship
management. These include Germany's FoodPanda (2019) operating in 43 countries and connecting with over 40,000 local restaurants; Swiggy (2019), India's
number one food app, downloaded no less than ten million times; UberEats (nd)
in 1,000 cities across the Americas and Asia; and GrubHub (2017) with 30,000
restaurants servicing over 800 US cities alone (Singh, 2018). This large range of
mobile food-related applications extends to online restaurant review platforms
such as Zomato (2018); Buycott (2018), which scans barcodes to provide information on product ethics; and Fooducate (2019), an ingredient substitute search
engine that 'empowers you to meet your diet, health and fitness goals' (Choi &
Graham, 2014).

These 'tactics of well-being' (Choi, 2014) are complemented by social media
platforms that share the aim of transforming the food system through connecting
rural food producers and urban eaters. They represent a move towards 'local,

community-driven and sustainable approaches to food and developments in social and mobile forms of technology that involve trust, sociality and network-logic' (Hearn, Collie, Lyle, Choi, & Foth, 2014, 211; Lewis, 2018a, 2018b; Lupton, 2018). These initiatives include Foodmunity (2011), an early attempt at 'connecting neighbours in a community through food', and i8dat (nd), an Instagram hashtag which facilitates the sharing of recipes and provides opportunities for interaction over food including 'celebration of diversity' (Hearn et al., 2014). Twitter, SMS, blogs and smartphone apps including Seasons (nd) and the late AgLocal (2017) are other examples of digital technologies engaged in 'material systems of acquisition' (Hearn et al., 2014, 206). These inform and connect food producers and eaters on multiple scales, while communities of growers are connected through Permablitz (2014), Transition Network (2016), Permaculture Worldwide Network (2017), Landshare (nd), Farmhack (nd) and Hyperlocavore (nd), some of which are 'yard-sharing', peer-to-peer learning networks that facilitate trade, exchange and borrowing of seeds, tools, knowledge, produce and even land.

It is challenging to identify how these proliferating consumer apps and platforms for 'non-institutional political participation' are interconnected. Do practices such as signing an e-petition, purchasing organic food and taking part in local urban farming initiatives interrelate, and if so, how and to what end (Witterhold, 2018) In this chapter, I argue that the time is ripe to consider how the 'quiet colonisation' (Lewis, 2018a) of our foodways by digital technologies can be directed towards the achievement of *food justice*, defined as 'a transformation of the current food system, including but not limited to eliminating disparities and inequities' (Gottlieb & Joshi, 2013, ix). Capturing 'the struggle against racism, exploitation, and oppression taking place within the food system' (Hislop, 2014, p. 19), food justice underpins an 'action agenda' (Gottlieb & Joshi, 2013, p. 233) with 'crossover appeal' to wider social justice movements. This suggests that our ability to change our food environments is limited unless we also enlist for battle against the unjust and immensely powerful economic, social and political forces that shape our foodscapes – the same forces that impinge on our digital environments.

Importantly, a focus on digital food justice brings into focus the social, economic and political dimensions of food choices and the 'cultural and media grammars' (Goodman, 2016, p. 264) that dictate our diets through powerful identifiers such as 'local', 'alternative', 'healthy' and 'industrial' food, while simultaneously recognising the limits of 'civic-minded digital consumerism' as a 'potentially collective and transformative act' (Lewis, 2018b, p. 218). Further, as Melissa Caldwell (2018, p. 25) notes, digital technologies are potentially themselves 'exclusionary', either through cost of access to reliable network coverage and the 'outpacing' of 'inclusive universal design features' by new innovations.

As new forms of sociality are emerging along the food chain with the sharing of our everyday experiences through networked technologies, where are the connections being made with groups which are marginalised in foodscapes?

What are the potentials for true civic engagement beyond digital food activism (Schneider, Eli, Dolan, & Ulijaszek, 2018; Schneider, Eli, McLennan, Dolan, Lezaun, & Ulijaszek, 2017) and app-driven engagement or 'apptivism' and other forms of 'connected consumption' (Lewis, 2018b)? How can digital technologies be applied in addressing the deep structural inequalities that impact on individuals and communities experiencing food insecurity and diseases of malnutrition?

This chapter interrogates the inherent contradictions of local food production and consumption and the digital applications that connect growers and consumers outside the conventional or 'industrialised' food system on a community level, which operates as 'a privileged site for analysing real-world communicative activity and the machinations of democratic life and civic engagement' (Broad, 2016, p. 18). Digital platforms and applications that embrace new politics of consumption and forms of digital food activism often have limited relevance to those most food insecure. These eaters are often marginalised through a complex set of factors known collectively as social exclusion, defined as 'a lack of connectedness' that implicates the social and physical environments in which people live, characterised by 'limited support networks, inability to access the labour market, alienation from society and poorer educational outcomes', all of which compound disenfranchisement (Vinson, 2009, p. 7).

The value of voice

Like food insecurity, digital exclusion is a symptom of wider social exclusion which affects individuals experiencing a lack of digital skills, barriers to connective infrastructure and access to technology complicated by affective factors such as perceptions of the relevance of technology; personal motivations and confidence in using it; and concerns about the safety of digital practices (Measuring Australia's Digital Divide, 2018). The socially excluded are often left out of digital conversations, and are thus denied opportunities for open, interactive exchanges about their lived experience (Couldry, 2015; Mann, 2018). The concept of social exclusion captures social disengagement and disconnection from services, as well as economic exclusion (Saunders, Griffiths, & Naidoo, 2008). This distinction highlights that income is not the only indicator of social exclusion which encompasses social isolation, social stigma, and lack of capacity and opportunity to voice one's own narrative.

Recapturing voice in the design and governance of inclusive food systems is the only viable response to what Nick Couldry (2010, p. 1) calls the 'contemporary crisis of voice' which many social groups are experiencing across not only political and economic but also social and cultural domains. This crisis can be directly linked to the dominant discourse of neoliberalism, which 'operates with a view of economic life that does not value voice and imposes that view of economic life onto politics, via a reductive view of politics as the implementing of market function' (ibid., p. 2). According to Couldry, this reductive view 'evacuates entirely' the role of the social in political regulation of economics.

I extend this interpretation to the corporate capture of voice and value in our foodways and argue that to reclaim our intimate experiences of food and rebuild the connections severed by its commodification, we must start from our lived experience in the communities in which we procure, consume and, if we are able, produce food (Mann, 2019). Our attachments to place include our food histories and form our ways of seeing and telling our stories about food. Voice powers one's ability to 'give an account of oneself' in the form of a narrative (Butler, 2005). To deny this capacity to possess and share one's narrative is to 'deny her potential for voice … a basic dimension of human life' (Couldry, 2010, p. 7). Voice is socially grounded, requiring resources in the form of language and status. It is a form of reflexive agency through which we 'disclose ourselves as subjects' (Arendt, 1958, p. 193) and make sense of our lives (Cavarero, 2000). As an embodied process of articulation, voice involves speaking and listening and is therefore an 'act of attention that registers the uniqueness of the other's narrative' (Couldry, 2010, p. 9) which respects also the internal diversity or plurality in each voice. We all have many stories, embedded in multiple contexts.

To be effective tools in the struggle for food justice, digital platforms purporting to be inclusive must accommodate social difference and acknowledge the way power 'enters speech itself' (Young, 1996, p. 123). This entails supporting the articulation of voices from the periphery of the market economy and respecting the inherent 'locatedness' of the eating experience, in that our engagement with our foodscapes is a series of 'intensely localised food events that require real choices by real people' (Belasco, 2012, p. 3). It recognises that while technological and social innovation that mobilises 'diffused social resources' including creativity, skills, knowledge and entrepreneurship is vital in creating just food systems, 'nothing can happen without the direct and creative participation of the people involved' (Manzini, 2011, p. 103).

Food justice-related themes of participation, equity and choice can be promoted by disruptive digital initiatives by 'changing the relationship presumed to inhere between problems and solutions' (Caldwell, 2018, p. 26). Food-centred and otherwise, activities that embrace principles of equity and participation to *transform* rather than merely *change* the food system, can contribute to addressing critical food justice issues (ibid.). Bringing people together and enabling them to identify and address their own concerns and find their own solutions can be a first step in the recovery and repair that is necessary to promote social inclusion and in doing so reduce social exclusion *and* rates of food insecurity. This demands a communication ecology perspective that focuses on change-making capacity at the community level, as 'a site where the challenges and risks of global society are experienced and addressed' (Broad, 2016, p. 18).

Connections and tensions

Digitally mediated food practices, like all digital practices, are 'unsettling other dimensions of daily life by introducing new forms of distance, alienation, or even invisibility' (Caldwell, 2018, p. 29). The local food movement, which has capitalised on new forms of digital agility through open source technologies and Web 2.0 by connecting producers and eaters in shorter supply chains, shares some of these tensions. Doing local food is about reconnection with physical spaces of food production but also, importantly, understanding the *what* and *why* of local food in communities. It is widely recognised as 'a collaborative effort to build more locally based, self-reliant food economies – one in which sustainable food production, processing, distribution, and consumption [are] integrated to enhance the economic, environmental, and local health of a particular place' (Feenstra, 2002). Valuing local foodscapes includes acknowledgement of how food enterprises strengthen local economics and emphasise self-sufficiency; how they contribute to community-building and collective action; and their potential to promote collective ownership, social inclusion, community empowerment and collective decision-making around a range of issues such as sustainable consumption.

Yet the tendency to assume something inherently positive about local foodways – the 'local trap' (Born & Purcell, 2006) – has been comprehensively critiqued in terms of leading to sustainability and social just outcomes. Further, discourses of localism are frequently co-opted. In Sydney, Australia, for example, in response to customer feedback, major retailer Woolworths is rebranding itself as a local champion, 'bringing neighbourhood grocers and local products' such as Hellenic Pastry and Pepe Saya into the store and creating a 'rustic yet future-proof design with real community spirit' (Chung, 2018). The domination of these major retailers, whose power in global supply chains is ever-increasing (Chemnitz, Luig, & Schimpf, 2017), reduces competition by creating an uneven playing field in which smaller retailers, including digital ventures like AgLocal (2017), cannot compete. AgLocal's obituary reads: 'we saw (and still do see) the internet as the most efficient way to connect people and build marketplaces' but having 'partnered with hundreds of farms across the country to tell the world the very human story of their work' and having 'built lifelong friendships with the families that support themselves via farming, we weren't able to reach our goals of becoming a business that can operate in a self-sustaining manner' (ibid.).

Other examples of online alternative food networks (AFNs) that operate within local foodshed boundaries aspire to 'strengthen connections to place and the local' while also creating global links, as in the case of *Open Food Network* (cited in Lewis, 2018a, p. 190). This type of online food network offers 'opportunities for reconnection with the "complex systems of food provisioning" that have worked to distance and disconnect consumers for the people and places involved in contemporary food production' (Bos & Owen, cited in Lewis, 2018a, p. 190). Australian examples using digital media include the online organic delivery services CERES Fair Food (2019) in Melbourne and Foodconnect (2019) in Brisbane, a social enterprise founded in 2005 by an ex-dairy farmer who aims to 'democratise the food system'. Comparable international initiatives include The Farmer: The Organic Store (nd) based at Thakkar Farm on the River Page,

90km from Mumbai, India, which supplies consumers with produce, fresh milk and herbal remedies via home deliveries and a distribution centre in Crawford Market in the city. Thakkar Farm is promoted on its website as 'pesticide chemical free since 1992' and situated in an 'eco-sensitive zone' distant from industry. The website also features a virtual farm tour, an Organic Bulletin and a link to Whats App. Embodying similar philosophies are platforms such as Farmer Uncle (2017) in New Delhi, Radish Boya (nd) in Japan, Tallo Verde (2018) in Buenos Aires and *Club Organico* (2018) in Rio de Janeiro.

It is clear viewing these sites that accessing and purchasing healthy, organic and/or sustainably produced food is more convenient and affordable for the digitally enabled, reproducing the disparities that exist in AFNs in the offline environment. Research indicates that organic food is generally more expensive than conventionally grown food and is the privilege of more affluent consumers. In Australia participation in AFNs, as well as in the more developed farmers market sector, is limited to middle-class consumers. The lack of participation on the part of those who are less affluent highlight the relatively high cost of produce and the lack of convenience associated with these forms of direct marketing as barriers to their involvement (Rose, 2017). Conventional, conveniently located and well-stocked supermarkets remain important for those consumers who may not have the time or means of transportation to purchase local food if it requires shopping at multiple or dispersed outlets (Hinrichs & Allen, 2008, p. 330).

Those who lack access to digital technologies or experience poor internet connectivity (such as rural broadband users in Australia) and those who live outside the distribution range of box delivery schemes or food-sharing networks are denied the option of participation. Correlations between limited access to fresh food and health-related disparities based on race, ethnicity and income are clear. Digital disparities replicate these patterns – the greater market penetration of fast food vendors in low-income areas without access to fresh, affordable food corresponds with a lack of delivery services for customers of online food food-sharing distribution app Ooooby (nd) in Sydney, for example.

Making local food the 'easy choice' for the mainstream consumer as well as the middle-class foodie is also critical in ensuring the economic viability of small-scale producers (James, 2016), many of whom come from Culturally and Linguistically Diverse (CALD) backgrounds and other marginalised and underserved populations, including lesbian, gay, bisexual, transgender or queer (LGBTQ), recent migrants and asylum seekers. For these people, participation in AFNs, small-scale food businesses and rural and urban farms can provide 'pathways to social empowerment, connection, community, and love' (Biel, LaScola, Berejnoi, Cloutier, & MacFayden, 2018). Measures to address exclusions include increased outreach and awareness of facilitating programmes, including food business incubators (Craven, Scholsberg, & Mann, 2018), and more inclusive, gender neutral and diverse terminology in different languages on platforms including websites; and further researching into the barriers to engagement in such programmes.

Participatory ecosystems

The barriers to participation in community food environments are becoming more porous within emergent, digitally supported food-sharing ecosystems (Edwards & Davies, 2018). This is particularly in cities where urban agriculture (UA), even if not playing a vital role in supplying food, is producing spaces that provide stronger connections to and understanding of the food environment, and in increasing awareness of the need to preserve small-scale and sustainable farming practices (Meenar & Hoover, 2012). The evolution of UA is propelled by 'ubiquitous technology, urban informatics and social media' (Hearn et al., 2014, p. 203) that facilitate the distribution and acquisition of food. For advocates, UA itself is a powerful connective force – it is about 'far more than growing vegetables on an empty lot. It's about revitalizing and transforming unused public spaces, connecting city residents with their neighborhoods in a new way and promoting healthier eating and living for everyone' (cited by Nordahl, 2014, p. 59).

UA is vitally important for developing food literacy and wider 'food vocabularies' that reflect understandings of culture and ethnic diversity in terms of 'which foods have meaning and value to the diverse racial groups that comprise our communities' (ibid, p. 157). Rohit Kumar (2013) observes the emergence of a grassroots local food movement that is 'cultivating better racial relations because it calls attention to our interconnectedness as humans living together on one planet'. Community gardens in Scotland have been identified as sites where political participation manifests itself through 'a process of *learning* by being in the presence of *difference*' (Crossan et. al. cited in Caldwell, 2018, p. 38).

These forms of engagement might be described as forms of 'civic agriculture': locally organised systems of agricultural and food production where networks of producers and consumers are bound together by place and a common mission to build a community's problem-solving capacity (Lyson, 2005). In successful cases, food becomes a mediator in learning and 'eaters are not just consumers but social actors whose meaning-making depend on faith, gender, age, income, or kinship' (Grasseni, 2017). In these initiatives people can participate not only as consumers but also as contributors and producers who learn from each other, peer-to-peer, through their food-sharing practices. This bi-directionality recognises that 'knowledge of agriculture is a resource that can be shared from poor to rich as well as from rich to poor' (Blevis & Morse, 2009, p. 61). Food-sharing and growing ecosystems can increase capacities for local 'content-creation', on- and offline, where individuals can provide descriptions and information about culturally and ethnically specific foods and food practices. The face-to-face engagements they enable can potentially contribute to social inclusion by developing relationships through 'moments of informal and unexpected sharing' including, as examples, 'cultural exchange with people seeking asylum who gifted home-made produce accompanied by a story from home' and unexpected episodes of generosity where those with 'home harvests...would offer more plant clippings than their initial offerings' (Edwards & Davies, 2018, p. 13).

Digital platforms facilitating these engagements include Popup Patch (nd), an edible community garden created in Melbourne in 2008 that was promoted via various digital media platforms to effectively shape messages about permaculture and growing food in small spaces (De Solier, 2018). RipeNearMe (2019), a web platform launched in July 2013 in Adelaide, Australia, connects local people with nearby backyard growers. Locals can also subscribe to individual fruit trees/produce to receive notifications. Transforming urban spaces into productive foodscapes (Brown, 2013), RipeNearMe (2019) became a financially sustainable social enterprise and a wider-reaching local food platform. The founders have turned to crowdfunding, which has proved successful, with 332 supporters helping the enterprise reach its crowdfunding target of $25,000 (StartSomeGood, 2014). The RipeNearMe community extended to 20,000 users around the world by the following year (Baldassarre, 2015). A similar initiative is OLIO, a surplus food-sharing app that aims to connect neighbours and local businesses on the basis that 'the act of sharing leftover food can bring in new customers, reduce food waste and help business[es] "connect with your community"' (cited in Lewis, 2018a, p. 189). In 2019, OLIO operated in 48 countries, had 900,000 users and 30,000 volunteers. It has facilitated the sharing of 1.2 million portions of food (OLIO, 2019).

Yet framing the act of sharing leftover food as 'a potentially collective and transformative act' overlooks the structural injustices in the system as a whole, and, as Tania Lewis puts it 'deflect[s] from rather than contributes to changing the reality of global agri-business by offering a quick ethical fix or salve for guilty consumers from the Global North' (Lewis, 2018b, p. 218). Edwards and Davies' (2018) research into the food-sharing ecosystem of Melbourne which includes a diverse range of actors including RipeNear.me identifies both positive and negative features of food-sharing ecosystems. There exist, for example, 'bottlenecks of competition and opportunity' for funding and volunteers and also strong relationships between like-minded organisations in sharing cooking spaces (Edwards & Davies, 2018, p. 12).

The Open Table initiative (2016), which collects surplus food from food rescue organisations, demonstrates the importance of, and synergy between, on- and offline communication and support platforms. Focusing on mitigating food waste and 'conviviality rather than food security, with a core focus on overcoming social isolation' (ibid., p. 7) through its 'monthly feasts', Open Table 'offer[s] a safe place for people to return and eat a free healthy meal'. In 'building skills and social connections' they are contributing to 'an actualised basis for change in small but substantial ways' (ibid., p. 15). Open Table demonstrates that while digital technologies present new channels for relationships building, they can also provoke '[mis]trust and social exclusion' (ibid.) when, for example, users experience discomfort in inviting strangers to visit their homes to collect produce. Further, those seeking asylum, the aged and the socially disadvantaged have difficulty accessing online platforms, leading several organisations to maintain traditional offline means of communication including phone calls, leaflet drops, community notices and house visits.

These cases reveal how food-centred practices and activities that occur in the physical communal environment, while not 'substituted' by online communities, can certainly be 'enhanced' (Vallauri, 2014, p. 173). This calls for 'generative approaches capable of attending to the complexity of urban relations and their productive political attention' (Turner, 2017, p. 4). Opening spaces for conversations about doing food differently can be facilitated by innovative technologies, as in the case of Growbot Gardens, at the San Jose Museum of Science and Technology, where diverse groups are encouraged to vision the future of small-scale agriculture through robotics and artificial intelligence (AI). Food hackathons, which seek through play and experimentation to 'transform food and food experiences in ways that make them accessible to as many people as possible, especially people who might otherwise be excluded from the food system as it currently exists', embrace the ethos of the participatory commons and the anarchist proposition that 'no one person owns ideas or knowledge or solutions' (Caldwell, 2018, p. 37). Creating the conditions for genuine participation includes providing 'formal and informal processes and networks through which communities make decisions and attempt to solve problems' (cited in Holley, 2016, p. 10).

Food-centred activities are just one way of drawing individuals together, on- and offline, as a path to inclusion. An excellent example of *transmedia organising* – 'the creation of a narrative of social transformation across multiple media platforms' that links attention directly to 'concrete opportunities for action' (Costanza-Chock, cited in Broad, 2016, p. 25) – the 'Every One. Every Day.' (nd) Project in the East London borough of Barking and Dagenham has spawned 250 local community projects, many initiated by residents, ranging from batch cooking with neighbours to a 'bee school'. Firmly embedded in community through an 'assets-based approach' that begins with 'drawing together existing opportunities' digital techniques and tools that work at a city and country level aim to increase levels of democratic oversight. For example: government websites for transparency; enabling digital activism in the form of online petitions and campaigning apps; providing digital platforms for crowdfunding which involves citizens 'co-investing' in shared public resources; and promoting crowdsourcing with citizens – that is, looking for new ideas and solutions to challenging problems (Participatory City, 2016, p. 50).

On a neighbourhood level, participants can view opportunities to participate through digital platforms which collect participation and direct outcome data to measure collective impact of project activity including environmental and social benefits such as tons of food grown locally, square metres of land under cultivation, zero carbon journeys, hours spent in physical activity, and overall sense of happiness and wellbeing ('Every One. Every Day.', 2018). These platforms are increasingly useful in collecting and measuring the public metrics and other data required to drive and measure the success of the project.

Developed following the observation that some 'innovative citizen-led local projects were achieving inclusive participation', the participatory culture approach of 'Every One. Every Day.' is based on 'common denominator activities' such as fixing, trading, singing, eating, playing, sharing cooking, learning, growing and making – acts of 'co-producing something tangible as a group of equal peers' (Civic Systems Lab, 2017, p. 20). These low/no cost, low commitment and inclusive activities can be labelled 'disruptive' in that they have the capacity to 'transform food and food experiences in ways that make them accessible to as many people as possible, especially people who might otherwise be excluded from the food system as it currently exists' (Caldwell, 2018, p. 37). Supporting 'cooperative efforts of mutual support and inspiration' (ibid.), this approach may lay the foundations for successful community-based food justice organising that includes local storytelling – 'conversations that emerge from lived experience and historical realities, are rooted in place, and are grounded in community-based collaboration' (Broad, 2016, p. 26). These conversations can be facilitated by digital technologies, which can also play a key role in 'cultivat[ing] networked partnerships that provide programmatic and fiscal sustainability' (ibid.). Key functions of the inclusive and participatory digital platforms that can support these initiatives are horizontal communication between members of the public and ease of access to and prompt notification of participation opportunities.

Characterised by a theory of change that locates local food justice activism within broader struggles for social inclusion, this approach is project-oriented, promoting group communication and learning. Here, learning can be understood as a 'distributed process' of translation across spaces, according to Colin McFarlane (2011, p. 363). He invites us to consider the role of 'urban learning forums' in acknowledging the 'attached and cluttered' nature of the individual citizen and facilitating the reaching of 'provisional understanding' (p. 370). McFarlane draws on the work of Callon et al. (2009, p. 18) who describe 'hybrid' forums as sites where 'questions and problems taken up are addressed at different levels in a variety of domains' by a wide range of experts and lay persons as ways of capturing diverse knowledge and experience. These include sites that connect on- and offline advocacy and promote political engagement, empowerment and information transparency such as Combat Monsanto (2008) and that emerged following the nuclear disaster in Fukushima, Japan, when consumers were unsatisfied by the level of radiation and food safety standards set by the government and corporations in (Kimura, 2013).

Digital platforms that promote these networks and narratives 'exhibit a willingness and capacity to develop community-focused action into large-scale cultural and political transformation' (Broad, 2016, p. 26). They offer possibilities for 'progressive forms of learning between different constituencies' (McFarlane, 2011, p. 360) if barriers such as access, education and affordability are overcome. They can facilitate and sustain the emergence of 'networked publics': 'imagined collective[s] that emerge as a result of the intersection of people, technology and practice' (boyd, 2010, p. 39) that participate in and, ideally, co-design healthy, sustainable and inclusive foodscapes.

Conclusion

Food justice extends well beyond the food chain. It embraces struggles against discrimination, exploitation, oppression and social exclusion in food systems. The food-sharing ecosystems presented in this chapter demonstrate the capacity of eaters and growers to exploit the 'positive double link between grassroots users and technology' (Manzini, 2011, p. 103). Digital platforms can go further to help mediate the participation of the food insecure in the co-creation of healthy, sustainable and inclusive foodscapes beyond provisioning and food relief. 'What "the digital" offers is not necessarily new techniques or fora but rather new ways of asking questions, generating insights, and forging communities' (Caldwell, 2018, p. 39). The networking affordances of digital initiatives offer ways to 'trace the flow of narrative, knowledge, culture, and identity' implicated in the ethos and actions of community-based food justice advocacy (Broad, 2016, p. 30).

These initiatives must privilege voice and listening to meet the demands of food justice as an element of social inclusion. This is a wider challenge that demands we build strong foundations for digital inclusion. Web literacy means negotiating the 'three-step path' of exploring, building and participating online and reducing barriers to the creation of local content that currently exist (Surman, Gardner, & Ascher, 2014, p. 68). Focusing solely on access is likely to amplify existing patterns of exclusion if content creation is not maximised to 'enhance local knowledge and strengthen champions of change' (Schoemaker, 2014, p. 80). These champions exist in local communities where their influence is essential to the social learning and attainment of individual wellbeing that can raise the capacity of all community members to participate in the co-creation of local foodscapes that provide comfort and safety as well as nutritious and affordable food. Given that 'it is people ultimately who create, adopt, adapt and govern technology as well as grow, process and consume food' (Davies, 2014, p. 192), these goals should be paramount in shaping our digital food cultures.

Further reading

Alkon, A., & Agyeman, J. (2011). *Cultivating Food Justice: Race, Class, and Sustainability.* Cambridge, MA: MIT Press.

Barking and Dagenham Growth Commission. (2015). *No-one left behind: In pursuit of growth for the benefit of everyone.* Retrieved from www.lbbd.gov.uk/wpcontent/uploads/2015/11/No-one-left-behind-in-pursuit-of-growth-for-the-benefit-of-everyone.pdf

Foley, R.A. (2018). Marketing critical consumption: Cultivating conscious consumers or nurturing an alternative food network on Facebook? In T. Schneider, K. Eli, C. Dolan & S. Ulijaszek (Eds.), *Digital Food Activism* (pp. 110–129). Abingdon, Oxfordshire: Routledge.

Giraud, E. (2018). Displacement, 'failure' and friction: Tactical interventions in the communication ecologies of anti-capitalist food activism. In T. Schneider, K. Eli, C. Dolan & S. Ulijaszek (Eds.), *Digital Food Activism* (pp. 130–150). Abingdon, Oxfordshire: Routledge.

References

AgLocal. (2017). Retrieved from https://aglocal.com/

Arendt, H. (1958). *The Human Condition*, Chicago: Chicago University Press.

Baldassarre, C. (2015). *Adelaide startup Ripe Near Me connects backyard growers with consumers looking to cut down food miles*. Retrieved from www.startupdaily.net/2015/10/adelaide-startup-ripe-near-me-connects-backyard-growers-with-consumers-looking-to-cut-down-food-miles/

Belasco, W. (2012). *Meals to Come: A History of the Future of Food*. Berkeley: University of California Press.

Biel, B., LaScola, D., Berejnoi, E., Cloutier, S., & MacFayden, J. (2018). *Does inclusivity really matter? The importance of diversity and inclusion in farm-based internship programs*. Retrieved from www.thesolutionsjournal.com/article/inclusivity-really-matter-importance-diversity-inclusion-farm-based-internship-programs/

Blevis, E., & Morse, S. (2009). Food, dude. *Interactions, 16*(2), 58–62.

Born, B., & Purcell, M. (2006). Avoiding the local trap: Scale and food systems in planning research. *Journal of Planning Education and Research, 26*(2), 195–207.

boyd, d. (2010). Social network sites as networked publics: Affordances, dynamics and implications. In Z. Papacharissi (Ed.), *Networked Self: Identity, Community, and Culture on Social Network Sites* (pp. 39–58). New York: Routledge.

Broad, G.M. (2016). *Food Justice and Community Change*. Oakland, CA: University of California Press.

Brown,J.(2013).*Ripenearme:Amarketforbackyardharvests,cultivatingcommunity.*Retrievedfrom www.cultivatingcommunity.org.au/ripe-near-me-a-market-for-backyard-harvests/

Butler, J. (2005). *Giving an Account of Oneself*. New York: Fordham University Press.

Buycott. (2018). Retrieved from www.buycott.com/

Caldwell, M. (2018). Hacking the food system: Re-making technologies of food justice. In T. Schneider, K. Eli, C. Dolan, & S. Ulijaszek (Eds.), *Digital Food Activism* (pp. 25–42). Abingdon, Oxfordshire: Routledge.

Callon, M., Lascoumes, P., & Barthe, Y. (2009). *Acting in an Uncertain World. An Essay on Technical Democracy* (G. Burchell, trans.). Cambridge, MA: MIT Press.

Cavarero, A. (2000). *Relating Narratives*. London: Routledge.

CERES Fair Food. (2019). Retrieved from www.ceresfairfood.org.au/

Chemnitz, C., Luig, B., & Schimpf, M. (Eds.) (2017). *Agri-food atlas: Facts and figures about the corporations that control what we eat*. Retrieved https://th.boell.org/sites/default/files/agrifoodatlas2017_facts-and-figures-about-the-corporations-that-control-what-we-eat.pdf

Choi, J.H. (2014). Tactics of well-being: Mobile media and a new turn in the human-food relationship. In G. Goggin and L. Hjorth (Eds.), *Routledge Companion to Mobile Media* (pp. 385–395). New York: Routledge.

Choi, J.H., & Graham, M. (2014). Urban food futures: ICTs and opportunities. *Futures, 62*, 151–154.

Chung, F. (2018). Woolworths unveils new-look flagship store featuring 'living' lettuce, meal kits and $14 chooks. Retrieved from www.news.com.au/finance/business/retail/woolworths-unveils-newlook-flagship-store-featuring-living-lettuce-meal-kits-and-14-chooks/news-story/63778db4a42adb1a2cc75723b921bc6e

Civic Systems Lab. (2017). *Designed to scale: Mass participation to build resilient neighbourhoods*. Retrieved from https://drive.google.com/file/d/0B28SOnHQM5HVV0pyT2p1NGNvQk0/view

Club Organico (2018). Retrieved from Radish Boya https://www.radishbo-ya.co.jp/shop/

Combat Monsanto. (2008). Retrieved from www.combat-monsanto.co.uk/

Couldry, N. (2010). *Why Voice Matters: Culture and Politics after Neoliberalism*. London: Sage.

Couldry, N. (2015). The myth of 'us': Digital networks, political change and the production of collectivity. *Information, Communication & Society, 18*(6), 608–626.

Craven, L., Scholsberg, D., & Mann, A. (2018). Addressing the SDGs through food business incubation: FoodLab Sydney. In *Food & Cities: The Role of Cities for Achieving the Sustainble Development Goals* (pp. 145–151). Retrieved from www.barillacfn.com/media/material/food_cities.pdf

Davies, A. (2014). Co-creating sustainable eating futures: Technology, ICT and citizen consumer ambivalence. *Futures, 62*, 181–193.

De Solier, I. (2018). Tasting the digital: New food media. In K. Lebesco & P. Naccarato (Eds.), *The Bloomsbury Handbook of Food and Popular Culture* (pp. 54–65). London: Bloomsbury Academic.

Edwards, F., & Davies, A. (2018). *Connective consumptions: Mapping Melbourne's food sharing ecosystem.Urban Policy and Research, 36*(4), 476–495.

Every One. Every Day. (2018). [PowerPoint slides] Retrieved from https://drive.google.com/file/d/1K6xFXRLTXIedjmtI4vmv7a4eO7Mkc-7W/view

Every One. Every Day. (nd). Retrieved from www.weareeveryone.org/

Farmer: The Organic Store, The. (nd). Retrieved from https://thefarmeronline.com//home

FarmerUncle. (2017). Retrieved from https://farmeruncle.com/

Farmhack. (nd). Retrieved from http://farmhack.org/tools

Feenstra, G. (2002). Creating space for sustainable food systems: Lessons from the field. *Agriculture and human values*. Retrieved from www.springerlink.com/index/G7887731X0263J18.

Foodconnect. (2019). Retrieved from https://foodconnect.com.au/

Foodmunity. (2011). Retrieved from https://austintoombs.com/foodmunity/

FoodPanda. (2019). Retrieved from www.foodora.com/

Fooducate. (2019). Retrieved from www.fooducate.com/

Goodman, M. (2016). Food geographies I: Relational foodscapes and the busy-ness of being more-than-food. *Progress in Human Geography, 40*(2), 257–266.

Gottlieb, R., & Joshi, A. (2013). *Food justice*. Boston, MA: MIT Press.

Grassini, C. (2017). *Food Citizens? Collective food procurement in European cities: Solidarity and diversity, skills and scale*. Retrieved from www.universiteitleiden.nl/en/research/research-projects/social-and-behavioural-sciences/food-citizens-collective-food-procurement-in-european-cities

GrubHub. (2017). Retrieved from https://grubhubmedia2015.q4web.com/media/overview/default.aspx

Hearn, G., Collie, N., Lyle, P., Choi, J., & Foth, M. (2014). Using communicative ecology theory to scope the emerging role of social media in the evolution of urban food systems. *Futures, 62*, 202–212.

Holley, K. (2016). *Equitable and Inclusive Civic Engagement: A Guide to Transformative Change*. Columbus: Ohio State University.

Hinrichs, C., & Allen, P. (2008) Selective patronage and social justice: Local food consumer campaigns in historical context. *Journal of Agricultural and Environmental Ethics, 21*(4), 329–352.

Hislop, R. (2014). *Reaping equity across the USA: FJ organisations observed at the national scale* (Master's thesis, University of California-Davis).

Hyperlocavore. (nd). Retrieved from https://hyperlocavore.wordpress.com/

i8dat. (nd). Retrieved from www.instagram.com/explore/tags/i8dat/

James, S.W. (2016). Beyond 'local' food: How supermarkets and consumer choice affect the economic viability of small-scale family farms in Sydney, Australia. *Area, 48*(1), 103–110.

Kimura, A.H. (2013). Standards as hybrid forum: Comparison of the Post-Fukishima radiation standards by a consumer cooperative, the private sector and the Japanese government. *International Journal of Sociology of Food & Agriculture, 20*(1), 11–29.

Kumar, R. (2013). *How the local food movement is transforming race relations in America.* Retrieved from www.huffingtonpost.com/rohit-kumar/transforming-race-relations_b_3140993.html

Landshare. (nd). Retrieved from www.landshare.org/

Lewis, T. (2018a). Food politics in a digital era. In T. Schneider, K. Eli, C. Dolan & S. Ulijaszek (Eds.), *Digital Food Activism* (pp. 185–202). Abindgdon: Routledge.

Lewis, T. (2018b). Digital food: From paddock to platform. *Communication Research and Practice, 4*(3), 212–228.

Lewis, T., & Phillipov, M. (2018). Food/media: Eating, cooking and provisioning in a digital world. *Communication Research & Practice, 4*(3), 207–211.

Lupton, D. (2018). Cooking, eating, uploading: Digital food cultures. In K. Lebesco & P. Naccarato (Eds.), *The Handbook of Food and Popular Culture* (pp. 66–79). London: Bloomsbury.

Lyson, T. (2005). Civic agriculture and community problem solving. *Culture, Agriculture, Food and Environment, 27*(2), 92–98.

Mackendrick, N. (2014). Foodscape, Key concepts in social research. *Contexts*, 6–18.

Mann, A. (2018). Hashtag activism and the right to food in Australia. In T. Schneider, K. Eli, C. Dolan, & S. Ulijaszek (Eds.), *Digital Food Activism* (pp. 168–184). Abindgdon: Routledge.

Mann, A. (2019). *Voice and Participation in Global Food Politics.* Abingdon, Oxfordshire: Routledge.

Manzini, E. (2011). The new way of the future: Small, local, open and connected. *Social Space*, 100–105. Retrieved from https://pdfs.semanticscholar.org/2dce/b9b5ba8293a530ed0de01ea726afed648cc1.pdf

McFarlane, C. (2011). *Learning the City: Knowledge and Translocal Assemblage.* Chichester: Wiley-Blackwell.

Measuring Australia's Digital Divide (2018). Retrieved from https://digitalinclusionindex.org.au/wp-content/uploads/2018/08/Australian-digital-inclusion-index-2018.pdf

Meenar, M., & Hoover, B. (2012). Community food security via urban agriculture: Understanding people, place, economy and accessibility from a food justice perspective. *Journal of Agriculture, Food Systems, and Community Development, 3*(1), 143–160.

Nordahl, D. (2014). *Public Produce.* Washington, DC: Island Press.

OLIO. (2019). Press Pack. Retrieved from https://olioex.com/about/

Ooooby. (nd). Retrieved from www.ooooby.org/sydney

Open Table. (2016). Retrieved from www.open-table.org/

Participatory City. (2016). *The Illustrated Guide to Participatory City.* Retrieved from https://issuu.com/participatorycity/docs/illustrated_guide_-_issuu_version

Permablitz. (2014). Retrieved from https://permablitzsydney.org/

Permaculture Worldwide Network. (2017). Retrieved from https://permacultureglobal.org/

Pop Up Patch (nd). Retrieved from https://littleveggiepatchco.com.au/pages/pop-up-patch

RadishBoya. (nd). Retrieved from www.radishbo-ya.co.jp/shop/

RipeNearMe. (2019). Retrieved from www.ripenear.me/

Rose, N. (2017). Community food hubs: An economic and social justice model for regional Australia? *Rural Society*, 26(3), 225–237.

Saunders, P., Griffiths, M., & Naidoo, Y. (2008). Towards new indicators of disadvantage: Deprivation and social exclusion in Australia.*Australian Journal of Social Issues*, *43*(2), 175–194.

Schneider, T., Eli, K., Dolan, C., & Ulijaszek, S. (Eds.). (2018). *Digital Food Activism*. Abingdon, Oxfordshire: Routledge.

Schneider, T., Eli, K., McLennan, A., Dolan, C., Lezaun, J., & Ulijaszek, S. (2017). Governance by campaign: The co-constitution of food issues, publics, and expertise thorough new information and communication technologies. *Information, Communication & Society*, *14*(6), 770–799.

Schoemaker, E. (2014). The mobile web: Amplifying, but not creating, changemakers. *innovations: Technology, Governance, Globalisation*, *9*(3/4), 79–90.

Seasons. (nd). Retrieved from www.seasonsapp.com/

Singh, A. (2018). Top 10 successful food delivery apps in the world. Retrieved from www. netsolutions.com/insights/top-10-successful-food-delivery-apps-in-the-world/

StartSomeGood. (2014). *Help establish an ultra-local food system*. Retrieved from https:// startsomegood.com/ripenearme

Surman, M., Gardner, C., & Ascher, D. (2014). Local content, smartphones and digital inclusion: Will the next billion consumers also be contributors to the Mobile Web? *innovations: Technology, Goverance, Globalisation*, 67–78.

Swiggy. (2019). Retrieved from www.swiggy.com/

Tallo Verde. (2018). Retrieved from www.talloverde.com/

Transition Network. (2016). Retrieved from https://transitionnetwork.org/

Turner, B. (2017). *Taste, Waste and the New Materiality of Food*. Abingdon: Routledge.

UberEats. (nd). Retrieved from https://about.ubereats.com/

Vallauri, U. (2014). Transition Belsize veg bag scheme: The role of ICTs in enabling new voices and community alliances around local food production and consumption. *Futures*, *62*, 173–180.

Vinson, T. (2009). *Social inclusion: The origins, meaning, definitions and economic implications of the concepet of inclusion/exclusion*. Canberra: Paper prepared for the Australian Department of Education, Employment and Workplace Relations (DEEWR).

Witterhold, K. (2018). Political consumers as digital food activists? The role of food in the digitisation of political consumption. In T. Schneider, K. Eli, C. Dolan, & S. Ulijaszek (Eds.), *Digital Food Activism* (pp. 89–109). Abingdon: Routledge.

Young, I.M. (1996). Communication and the Other: Beyond deliberative democracy. In S. Benhabib (Ed.), *Democracy and Difference: Contesting the Boundaries of the Political* (pp. 121–135). Princeton, NJ: Princeton University Press.

Zomato (2018). Retrieved from www.zomato.com/sydney/restaurants

11

VISIONING FOOD AND COMMUNITY THROUGH THE LENS OF SOCIAL MEDIA

Karen Cross

Introduction

Recent media scholarship has placed particular emphasis on the elision of previously held distinctions between the amateur and the professional. It has further documented the development of 'produsage' (Bruns, 2008) and 'prosumption' (Fuchs, 2013) cultures and their role in sustaining new performances of labour online. Scholars have also begun to draw attention to this within the context of digital food networks and the development of social media platforms, reflecting especially on new entrepreneurial strategies of self-promotion and branding (see, for instance, Rousseau, 2012). Within this, they have also considered the development of new forms of creative labour (De Solier, 2018) that are on display when digital media and the production and consumption of food come together. Researchers have also considered the ways in which new digital food cultures initiate forms of (self)surveillance, bodily and aesthetic control (Lewis, 2019; Lupton, 2017), and paid attention to the new commercial imperatives created by sharing of food-related data online (Lupton, 2018a, 2018b).

The biopolitical forces of food mediation are now shown to have particular consequences for society and the individual as they navigate new terrains of food mediation (Goodman, Johnston, & Cairns, 2017). Work in the field of food studies has also begun mapping the growing popularity of alternative food politics (Phillipov, 2019) and the role of digital technologies in enabling the growth of food activism (Schneider, Eli, Dolan, & Ulijaszek, 2018). This research draws attention to the importance of the concept of 'space' within food mediation, especially the geolocative strategies of 'virtual reconnection' (Bos & Owen, 2016) now made available in online platforms. Linked to this, Tanya Lewis' (2018a, 2018b) more recent work on food and social media has also usefully drawn attention to the role of photography and other visual media within the development of digital food networks by focusing on the transfer of food politics 'from paddock to platform' (Lewis, 2018a).

While it is clear that food media are entangled in a complex web of privatised and individualistic market orientations, emplaced communities also remains an important part of the picture of digital food mediation. As I aim to show in this chapter, community has become increasingly important because it activates a sense of social connection in online networks. However, it is also the case that actual communities can be impacted by the different performances of connectivity that have become associated with digital food media, and the rise of the network and social media visualities with which it is associated.

Questions that arise are:

- How do social media networks mediate community?
- What is the role of food within this, and how does food consumption and performances of alternative food politics in particular form an important aspect of community making today?
- How does the new spatial dimensions of the online network (a largely urban construct) underpin the making of community today?
- In what ways does social media become the lens through which food and community are experienced and understood?
- How does digital food media impact upon communities and their future development?

In order to be able to respond to these questions, it is important to bear in mind the wider picture, or ecosystem, within which the mediation of food and community takes place. There are various approaches that scholars have employed in recent years, including one described by De Solier (2018, p. 54; see also De Solier, 2013) which involves investigating food media as a genre of 'material media'. That is to say, food media can be considered a part of a landscape of other material culture that 'are central to how objects are used in post-industrial lifestyles and self-formation' (de Solier, 2018, p. 55). However, we may also want to think about how so-called 'material media' are entwined with the forces of material mediation that are also made apparent by digital culture.

Another way to understand this is by considering the interchange between the analogue and the digital and the way that interactions between them are sustained within online spaces. Instagram could be taken as an obvious example here, as it privileges the past of photography: the mundane and familial frame of the personal snapshot. This plays upon the perceived notions of authenticity and memory, which helps in furthering the goal of connectivity. It is also important within the discussion of new media aesthetics to consider the role of space and place-making (both as analogue and as digital constructs) pertaining to community contexts.

To locate this discussion, I begin with a brief critical reflection of the notion of 'the networked society' as proposed by Manuel Castells (1996) and the dominance of urban interests within the political forms of 'grassrooting' that he describes. I then explore the role of digitalised forms of 'space making' in the growth and development of alternative practices of food consumption within the setting of the urban high street in south London. Two sites form the central focus here. The first is the redevelopment of Brixton Village Market by a self-described 'utopian regeneration agency' known as Space Makers, established in 2009. The second is that of a volunteer-led community project known as the West Norwood Feast, established in 2011 with the support of the Space Makers agency. In relation to these two ventures, the chapter traces the impact of online networks and social media within the development of community-based food markets. Finally, it reflects upon the wider impact of social media upon the development and regeneration of urban communities and their linked performances of food consumption.

Digital urban food imaginaries and the idea of 'the network'

The spatialising effects of web-based communications have led to the accepted dominance of urban imaginaries within the development of digital cultures. The writings of Manual Castells and his idea of 'the network society' (1996) has been especially influential in the development of critical thinking in this area. Very early on in his book *The Informational City* (1989), Castells makes a general distinction between 'the space of places' (a more traditional conception of physical location) and 'the space of flows' (a connective horizon of experience liberated from traditional place/time constraints). As Castells recognises in his later work, however, it is in fact the interrelation between these two planes of 'places' and 'flows' that enables the 'grassrooting' (1999, p. 294) effect that is required for social change. As he writes:

> [T]he geography of the new history will not be made of the separation between places and flows, but out of the interface between places and flows and between cultures and social interests, both in the space of flows and the space of places.
>
> *(1999, p. 294)*

This suggests that that 'the network' is multidimensional in nature and is in fact that which is formed in the dynamic interaction and transactions between the on- and offline worlds, between cultural and societal interests. However, for Castells 'the network' is also that which is fundamentally premised upon the infrastructure of the city and is inherently guided by its systems of conveyance and transfer; for example, the transport lines and links that enable the mobility of information and people. The notable absence here is that of the rural landscape, especially the sense of rootedness that is gained from a connection with the land. Furthermore, the role of less obviously geographic spatialities should be considered, such as those that feature within new networked urban imaginaries and within the emergence of 'material media' I previously described.

Considerable critical attention has been paid within the urban studies literature to the forces of privatisation and individualism that become apparent within
the city. Discussions that reactivate Henri Lefevre's (1968) notion of 'the right
to the city' have been especially forceful. A notable example is David Harvey's
work on *Rebel Cities* (2012, p. 4), in which he argues that 'reinventing the city
inevitably depends upon the exercise of collective power over the processes of
urbanization'. However, there is also a need to be attendant to the apparently
more mundane socialities of the city that, as I demonstrate in this chapter, link
together the politics of space occupation with the making of alternative food
markets.

As Myria Georgiou (2013, p. 6) has recently highlighted in her work exploring new processes of cosmopolitanisation linked with the city, it is important to
identify the role of media and new processes of digital mediation that enable new
citizenry identities and democratic forms of participation that ultimately support
and sustain private capital accumulation. In her work on food markets in East
London, she explores how new global strategies of participation are often staged
on the basis of inherently aesthetic interests and commercial imperatives imbricated in a politics of class and race. People choose between a range of options
within the market place, and, moreover, represent these choices through new
forms of mediated sociality, especially in online social media platforms. These
displays represent the embodied and pleasurable experiences of consumption,
and the making and sharing of photographs that document a sense of togeth-
erness enhances the sociality of these performances. However, certain rules of
signification apply and these are importantly connected to questions of social
representation and 'the right to the city' as well as raising questions of food
justice.

Snapshots of food posted online can be enactments of 'embodied intimacy'
(Lewis, 2018a, p. 213) but it is important to recognise also that these are related
to the politics of gender, class, race, ethnicity and the vast range of cultural
nuances that come to play within the context of the mediated performances of
eating. The social connection promoted within the context of digitally mediated
performances of food consumption is not always widely available, nor necessarily
desired by all. Furthermore, the consequences of new technologies for communities need to be born in mind, as now I turn to consider in relation to the case
of Brixton Village Market, the first site that forms the central focus of this study.

The politics of space making in Brixton Village Market

Granville Arcade is located in Brixton Village Market in south London and has
deep historical links with the *Windrush* generation, a community of predominantly Caribbean people who migrated to the UK during the post-war era. The
market was once a thriving shopping and entertainment area for this community
but since this time Brixton has become a problematic site of race relations. During the 1980s and 1990s it was impacted by the growth of the enterprise culture

that took over in government planning agendas (Deakin & Edwards, 1993). The success of communities became tied to the success of private corporations and processes of gentrification resulting from investment in commercial markets.

The downturn of the economy in more recent times has deepened the sense of social division but more recent processes of market development have also become tied to the new incoming forces of network activism. Brixton continues to be affected by gentrification processes. Due to its excellent transport links with the city and the redevelopment projects that it has seen over time, it has become increasingly appealing to a new younger, predominantly white, middle-class population. The growth of its alternative food markets has also made the area appealing. The market has become host to a regular farmer's market as well as a number of pop-up shops, new art galleries, cafes and bars. Brixton has been identified as a site of 'foodie' interest for 'the out-of-towners, drawn by the prospect of eating (and tweeting) at London's hippest food stalls' (Godwin, 2013). However, these newer aspects of urban consumption find their roots in specific activist interventions guided by the principles of the network.

During 2009, a group known as Space Makers led by the journalist and social entrepreneur Dougald Hine announced that it had agreed a proposal for the occupation of 20 shop units in the market, which halted a previously planned development to turn the market into high-rise flats with a supermarket outlet at the bottom. Driven by the ethics of the startup, the market initially took the form of a pop-up with a festival of youth groups and arts organisations supporting during the initial weeks (Shalet, 2010). Space Makers, although initially only represented by a group of like-minded individuals connecting online, soon formalised to pool their connections with key organisations, such as the Empty Shops Network and sustainable architectural organisations. In addition, Hine drew upon his contacts with NESTA (the National Endowment for Science, Technology and the Arts) and the Young Foundation, both charitable organisations interested in driving forward social change through investment, policy and research.

The Space Makers initiative attracted widespread positive press coverage (Hirschmiller, 2010; BBC News, 2011), which also included a feature in the *New York Times* (Wilder, 2010) suggesting that the market had brought a 'fresh face' to south London. Along with the West Norwood Feast, discussed later this chapter, it was also cited in Mary Portas' review into the future of Britain's high streets (Portas, 2011) as an example of a successful intervention. Yet it is also clear that for many local people, the market has come to form an unwanted intervention or aspect of 'greenlining' (Anguelovski, 2015) that researchers show is having particularly detrimental consequences for working-class communities. Furthermore, it also involves a specific politics of race commonly associated with community food markers, representing a form of 'whiteness' (Slocum, 2007, p. 520), which, on the one hand, functions as a 'hopeful vision of changing communities' but, on the other, assumes a white middle-class consumer base.

There is a great deal of ambivalence about what has happened in the case of Brixton with the Space Makers Agency (Brixton Buzz, 2018) and there is a distinct sense that the new stallholders who entered the market space through the redevelopment stand to represent a very different set of community values and interests. Whereas in the post-war period the market focused on bringing foods from abroad to its migrant community, today gourmet cafes and restaurants now reflect a different kind of social experience that, on the one hand, represents the 'bottom up energy' (Hine, 2009) required to counteract big high street monopolies, but, on the other, forms an homogenised environment of racialised consumption.

At the worst end of this, the market has been linked to the growth of the new urban class of 'champagne and fromage guzzling yuppies' (Lubbock, 2013) recalling ever more deeply the legacy of enterprise culture pervading UK planning policy. The arrival of the new urban elite comes as an unwelcome intervention that is especially contentious in light of Brixton's associations with the 2011 riots, which highlighted problems of social exclusion impacting upon young black men especially. At a later protest in the market in 2013, one local described 'living in London as like starving to death at a feast' (Lubbock, 2013) but the new divisions of class and social uprising are performed along complex lines of cultural resistance tied into the counter-politics of the networked society.

More recently, in 2015, Brixton Village Market became the setting for local protests that were linked to the Black Lives Matter movement, with some demonstrators displaying banners using the #blackcommunitiesmatter hashtag and calls to recognise the importance of the working-class community and the specific legacy of resistance historically associated with the area. This came at a moment in time when Brixton market was also beginning to see the infiltration of more aggressive corporatism. A series of rent rises had already occurred, and a sense of distrust between the market traders and landlords was growing (Bryant, 2013).

The market has been sold to a property investment and asset management company known as Hondo Enterprises, which is owned by a Texan socialite with links to the royal family (Brixton Buzz, 2018). Shop units that once sat within the arches of the linked railway line have faced a programme of mass eviction due to Network Rail's national scheme of selling property for further investment. These have previously played an important role in aiding the development of small businesses and in sustaining the sense of diversity in local areas (Brett, 2018), but it is increasingly apparent that the cultural specificities of the market and those whose lives are interconnected are being pushed out.

Network politics and the aesthetics of social media

In the introduction to a recent edited collection *Alternative Food Politics: From the Margins to the Mainstream*, Michelle Phillipov (2019, p. 2) raises a series of important questions, including asking: 'how do we conceptualise the cultural work that media does in shaping contemporary food politics? Where is power located in these new assemblages of media and activist and market forces?' One way to respond is to investigate the particular space-making practices that proliferate through contemporary network imaginaries. As the case of Brixton shows, we can trace the cultural effects of digital media practices and the investment in the discourse of the network through the rise of new market materialities and their associated performances of sociality.

As Phillipov (2019, p. 5) also discusses, lifestyle media now adapts elements of alternative food networks as new signifiers of desirability for the urban middle class. This in part echoes Kate Soper's notion of 'alternative hedonism' (2004, p. 112), where idyllic rural settings offer 'a site of rescue and purification from the ravages of corporate, urban life' but within the context of digital performances of lifestyle new political challenges arise. The use of digital platforms can effect change in food systems and shift dominant discourses of food production (Schnieder, 2019). However, we also need to understand digital media's role in relation to communities and what support they can play not just in terms of environmental sustainability but also in alleviating social inequalities and social marginalisation.

There are specific consequences that arise from the kinds of network interests described here as they spill over onto high streets and other market spaces. It may not be easy for existing members of the community to identify with these, and the tensions that have become apparent in Brixton reveal how the problems of class and race persist within new digital social configurations of food consumption. It can also be said that alternative food markets and the kinds of consumption they promote are clearly indivisible from questions of property ownership. Equally so, the case of Brixton shows that it is important to challenge the role played by network actors who make claims to the city in the name of social change.

Although emerging to support small businesses and diversity in the market, pop-ups have a particular role to play in terms of establishing new forms of community space. As Mara Ferreri (2016, p. 141) writes, pop-ups function as 'urban interruptions' that enable strange and spectacular, yet domestic and familial, encounters within the space of the high street. The do-it-yourself aesthetic provides a creative filler and sense of communal festivity in an otherwise predictable and humdrum flow of market exchange. Pop-ups are reflective of the 'bottom up energy' previously described, but they are also now considered desirable in venture capitalist contexts of urban development.

New practices of alternative food consumption can be said to support the 'new moralities of productive leisure' (De Solier, 2013, p. 20) and process of 'self-formation' (De Solier, 2018, p. 58) that predominate within visual social media. It is in displaying our specific acts of consumption within the market that we also display our capacity to decipher the new codes and conventions of alternative social enterprising identities more generally. Geolocation activities form a central feature of new leisure-time mediality. Identification and use of certain hashtags are important, but so too is the correct use of the 'right' platforms and the particular modalities of communication that they support.

As De Solier (2018, p. 62) has noted in relation to her work on urban gardening, there are different kinds of culinary capital associated with different platforms. YouTube, for instance, can be recognised as an overtly more practical educational frame, whereas other visual media focus more overtly on aesthetic displays. However, while the different registers of media are important to consider, it is equally important to account for the visual aesthetic codes that are associated with food media. In relation to this, Tanya Lewis argues that photographic depictions of food consumption often involve creative artisanal performances, which become recognisable within the network by their 'double mastery of the crafting of food *and* the shaping of associated imagery' (2018a, p. 214, emphasis in the original).

Extending on from this, it is possible to also argue that the craft dimensions of 'foodie' culture can be linked rise of interest in retro culture visualities and desire for the past on display in social photographic cultures (Cross, 2018b). This is especially apparent within Instagram and the individualised self-promotional performances this platform tends to favour. It is here that we see aesthetics of the snapshot lens providing contemporary media users with the filtering effects that serve to enhance sense of authenticity and intimacy within performances of community food. As I now turn to describe, this represents an important extension of digital networked materialities within the material spaces and practices of community markets.

West Norwood Feast

The Space Makers founder Dougald Hine now lives in rural Sweden and focuses his attention on ways of unplugging from the seductions of screens and networks (Hine, 2018; Smith, 2014). The Dark Mountain project, which he began at the same time as working on Brixton Village, provides a platform for more radical ideas of sustainability that involve returning to the wilderness. In contrast to the startup ethos of previous years, the preference now is for more deeply rooted forms of connection to the land, expressed also in the contemporary development of alternative food politics. However, the effects of Hines' earlier network-inspired interventions have had a lasting impact upon the local community and in turn continue to influence the direction of local policy and neighbourhood planning strategies.

Reflecting on his previous work with a newfound wisdom, Hine (2018) notes that he is prouder of the West Norwood Feast initiative than he is of the Brixton Market project, which is the one that got all the attention. The Feast was formed following on from a series of meetings led by Space Makers in a local arts centre in 2011. A hand-drawn map that is displayed on the website https://westnorwoodfeast.com/history/ tells the story of the market's evolution, and by placing key highlight points such as 'like minds' and 'genuine collaboration' over the different parts of the visualisation, the suggestion is that this was a project that involved the whole community and was focused on gaining 'trust from local stakeholders' and achieving a sense of 'pride of place'.

As the annotations express, the aim was also to provide an 'antidote to a depressing high street' with the inclusion of key areas of consumption including an artisans market, street food, a retro market and family hub, again playing up the defining characteristics of community-based pop-ups. The market has achieved significant success and continues to be hosted on a monthly basis during the months of April to December. Frequently featured in *Timeout*, it stands as a marker of success in community development. The central food area, which is located on the green outside the local church, provides a home for music and entertainment groups and there is a strong focus on promoting community-led arts organisations. These also provide a sense of festivity that is important to its hedonistic celebrations of food and diversity.

Not in question here are the motivations of those who work hard volunteering to support the Feast, for clearly their intentions are to build a sense of community that is inclusive. The market stages free activities for children and for pensioners, thus representing those who have historically tended to be marginalised in cities. In tandem with its own events, it works with the library and local leisure centre to provide a programme of activities that may be of interest to the wider community. However, its main business is the promotion of alternative food markets and it requires a literacy of taste-making practices to make its experiences of consumption legible.

The do-it-yourself ethos replicates itself within this context. Signifiers of post-war domesticity, including the widespread consumption of home-baked cakes and use of bunting, feature widely across the market. The now familiar use of the 'keep calm' logo, suggesting austerity resilience, alongside signs that tell shoppers to 'keep it local' (see Figure 11.1) together imply the need for an investment in the home front. It is clearly not the shops that are located in the high street to which this sign refers. 'Local' does not mean the high street pharmacy or supermarket. Rather, it refers to the products made by artists and small businesses, which largely operate online when not in the market space.

The complicated race relations of Brixton Village Market are also apparent in West Norwood, and link to a broader issue of the connection between food politics and pro-localism. Drawing on the work of the geographer Doreen Massey (2005, p. 170), Rachel Slocum (2007, p. 531) suggests the need to attend to the questions of power that lie in forms of activism that promote pro-localism. There is a wider history of problematic state and institutional attempts to use food to actively leverage forms of national identity (see, for example, Pilcher, 2012). This is not something that necessarily disappears within the development of alternative markets just because they appear to represent 'grassroots' views.

The demand for ethical and sustainable alternatives also arguably necessitates an ethics of social justice, which can account for new formations of racialised food practice and the role of media within this. The uploaded images tagged by area enables the social media user to identify with the iconicity of certain spaces and their distinctive socialities of food consumption. These displays also allow the user to identify as a politically interested and ethical being. The paraphernalia

FIGURE 11.1   Keep it Local sign featured on @wn_feast Instagram page 2016

of the market, such as canvas tote bags displaying signs such as #WeNo (a shortened version of West Norwood mimicking Soho), People Powered and Community Rocks, further enhances the spectral performance of networked urban community politics. These are products and stagings of community made to be shown online, and they represent the now more common aspects of 'shopping for change' associated with the popular representations of food activism.

This is not to deny the important consequences of genuine collaboration that can arise in alternative networks, which are especially important for the environment (Alkon & Guthman, 2017). However, just as Lewis (2018b, p. 197) has recently argued, there is a need to not avoid the hard questions about the 'progressive participatory promise of media tools and infrastructure that are increasingly dominating our daily lives'. As digitally mediated signifiers of connected sociality proliferate online, and increasingly mediate our sense of food and community, there is a risk that the ability to engage in social politics and express citizenry interests is premised simply upon people's capacity to participate in spectacular displays of alternative consumption.

The tensions within the West Norwood community have become evident more recently in relation to the opening of a branch of well-known fast food chicken shop chain on the high street. The local neighbourhood Facebook forum has seen a number of discussion threads appear on the topic, and it is clear from the comments that food consumption is considered a particularly emotive topic for many in the community. The assumed preference for healthy alternatives has more recently become challenged and performances of taste in relation to this have become a means by which a new class politics and resistance to new market directions is being enacted. The defence of the high street chain allows local people who feel marginalised by middle-class preferences to undermine what they perceive to be the increasingly encroachment of an exclusive zone of middle-class 'white' consumption.

Conclusion

As in many UK cities, the gap between the rich and the poor in London is widening. The Norwood & Brixton Foodbank (2018) reported a 22 per cent increase in use during 2017–2018, with many of its users known to be in paid employment but struggling still to meet their own basic needs. The representation of working-class food politics becomes all the more pressing in this context. However, in situations where the market forms the baseline for the formation of community, and its future success, it is difficult to see how the needs of the wider community can be met.

Recent planning documents for the wider area (Lambeth Council, 2015) cite the Feast as a successful example of community regeneration. This is not surprising given the historical reliance upon technological futurisms within urban planning. The concepts of 'vision' and 'visioning' commonly used by planners around the world for a number of years (Shipley, 2000) take on new meaning in this context. They emphasise not only the occularcentricity of Western culture in the governance of social space but also now the importance of social imaginaries in communicating a sense of community.

Media and communication scholars are well positioned to understand the problems associated with the concept of 'visioning' (especially the representation of marginalised groups), but further research is now needed to establish a sense of the wider ways in which food mediations and related network identities provide a frame of reference for community-building activities. As the successes of such market developments seen in Brixton and in West Norwood become the benchmark for future planning visions, the combined critique of mediated food networks and the uses of city spaces is especially urgent.

As this chapter has sought to show, there is a risk that aspects of food justice of increasingly importance may become lost within the network politics governing urban community building. As discussed, the space-making practices of the network are partial and represent the dominance of the city and its technological infrastructure. While it is clear that community is never a fixed and stable entity, nor an unchanging site of class interest, the concerns of community do not just disappear with the rise of digital food cultures now dominant in the popular cultural visual landscape of social media.

The socialities of the digital network are slippery in terms of their market effects. While they can work to transform the terrain of politics, these socialities and networks require us to also seek out new horizons that can respond to and incorporate specific community concerns that continue to remain under- or misrepresented within the new space-making practices governing urban centres.

The future food highways, smart food cities and community building so celebrated in recent policy planning and urban strategy are currently found wanting. Images of food circulating online represent the new taste-making processes of networked identities and the aspects of self-making associated with online promotional frames. Visual social media has also become a key part of community-making and are important to its success, but exactly how networked urban imaginaries that form within and proliferate through social media are allowed to form the basis of future planning for communities needs further research.

Further reading

Goodman, M.K., & Sage, C. (2016) *Food Transgressions: Making Sense of Contemporary Food Politics*. London and New York: Routledge.
Lewis, T., & Potter, E. (Eds.). (2011) *Ethical Consumption: A Critical Introduction*. London and New York: Routledge.
Moe, B.W., & Shurance, K.R. (2017) No longer tied to the local: street food's technological revolution. In K. Lebesco & P. Naccarato (Eds.), *The Bloomsbury Handbook of Food and Popular Culture* (pp. 138–154). London and New York: Bloomsbury.

References

Alkon, A., & Guthman, J. (Eds.). (2017). *The New Food Activism: Opposition, Cooporation, and Collective Action*. Berkeley: University of California Press.
Anguelovski, I. (2015). Healthy food stores, greenlining and food gentrification: Contesting new forms of privilege, displacement and locally unwanted land uses in racially mixed neighbourhoods. *International Journal of Urban and Regional Research*, *39*(6), 1209–1230.
BBC News. (2011). *Breathing new life into Brixton market*. Retrieved from www.bbc.co.uk/news/av/business-16159294/breathing-new-life-into-brixton-market
Bos, E., & Owen, L. (2016). Virtual reconnection: The online spaces of alternative food networks in England. *Journal of Rural Studies*, *45*, 1–4.
Brett, W. (2018, August 15). Don't sell of the arches, Network Rail – it will kill off diverse business dreams. *The Guardian*. Retrieved from www.theguardian.com/commentisfree/2018/aug/15/network-rail-arches-small-business-sell-off
Brixton Buzz. (2018). *Brixton fifteen years ago*. Retrieved from www.brixtonbuzz.com/2018/02/brixton-fifteen-years-ago-brockwell-snow-brixton-village-popes-road-and-lido-photos-feb-2003/
Bruns, A. (2008). *Blogs, Wikipedia, Second Life and Beyond: From Production to Produsage*. New York: Peter Lang.
Bryant, M. (2013, January 3). Soaring rents are forcing us out, say Brixton market traders. *Evening Standard*. Retrieved from www.standard.co.uk/news/london/soaring-rents-are-forcing-us-out-say-brixton-market-traders-8436274.html
Castells, M. (1989). *The Informational City: Information Technology, Economic Restructuring, and the Urban-Regional Process*. Oxford: Blackwell.
Castells, M. (1996). *The Rise of the Network Society. Vol. 1 of the Informational Age: Economy Society and Culture*. Cambridge, MA and Oxford: Blackwell.

Castells, M. (1999). Grassrooting the space of flows. *Urban Geography, 20*(4), 294–302.

Cross, K. (2018b). From visual discipline to love work: The feminising of photographic expertise in the age of social media. In S. Taylor & S. Luckman (Eds.), *The New Normal of Working Lives: Critical Studies in Contemporary Work and Employment* (pp. 65–86). Basingstoke: Palgrave Macmillan.

De Solier, I. (2013). *Food and the Self: Consumption, Production and Material Culture.* London: Bloomsbury.

De Solier, I. (2018). Tasting the digital: New food media. In K. LeBesco & P. Naccarato (Eds.), *The Handbook of Food and Popular Culture* (pp. 54–65). London: Bloomsbury.

Deakin, N., & Edwards, J. (1993). *The Enterprise Culture and the Inner City.* London and New York: Routledge.

Ferreri, M. (2016). Pop-up shops as interruptions in (post-)recessional London. In S. Jordan & C. Linder (Eds.), *Cities Interrupted: Visual Culture and Urban Space* (pp. 141–156). London and New York: Bloomsbury.

Fuchs, C. (2013). Digital presumption labour on social media in the context of the capitalist regime of time. *Time and Society, 23*(1), 97–123

Georgiou, M. (2013). *Media and the City: Cosmopolitanism and Difference.* Cambridge: Polity.

Godwin, R. (2013). Is gentrification killing Brixton Market? *Evening Standard,* 25 January. Retrieved from www.standard.co.uk/lifestyle/esmagazine/is-gentrification-killing-brixton-market-8465017.html

Goodman, M.K., Johnston, J., & Cairns, K. (2017). Food, media and space: The mediated biopolitics of eating. *Geoforum, 84*, 161–168.

Harvey, D. (2012). *Rebel Cities: From the Right to the City to the Urban Revolution.* London and New York: Verso.

Hine, D. (2009). *Spacemakers agency.* Retrieved from http://otherexcuses.blogspot.com/2009/09/space-makers-agency.html

Hine, D. (2018, March 12). The Long Journey HOME: On (Finally) Starting a School. *Medium.* Retrieved from https://medium.com/a-school-called-home/the-long-journey-home-on-finally-starting-a-school-360e356a9f

Hirschmiller, S. (2010, August 12). How Brixton's Arcade got a new lease of life. *Evening Standard.* Retrieved from www.standard.co.uk/lifestyle/how-brixtons-arcade-got-a-new-lease-of-life-6502388.html

Lambeth Council. (2015). *Invitation to tender for visioning West Norwood and Tulse Hill.* Retrieved from www.contractsfinder.service.gov.uk/Notice/downloadAttachment?noticeId=5f68dda8–3668–4018-a519–2d147d64a3f8&id=edfcf19b-2813–4b63–92c3–6415137929a2

Lefevre, H. (1968). The right to the city. In H. Lefevre, *Writings on Cities* (pp. 63–184). Cambridge, MA: Blackwell.

Lewis, T. (2018a). Digital food: From paddock to platform. *Communication Research and Practice, 4*(3), 212–228.

Lewis, T. (2018b). Food politics in the digital era. In T. Schneider, K. Eli, C. Dolan, & S. Ulijaszek (Eds.), *Digital Food Activism* (pp. 185–202). London: Routledge.

Lewis, T. (2019). Digital food: From paddock to platform. *Communication Research and Practice, 4*(3), 212–228.

Lubbock, J. (2013, October 14). Yuppies out protested Brixton's newest champagne bar on Friday. *Vice.* Retrieved from www.vice.com/en_uk/article/3bw8z3/yuppie-out-tried-to-kick-the-yuppies-out-of-brixton-by-drinking-cider

Lupton, D. (2017). Digital media and body weight, shape, and size: an introduction and review. *Fat Studies, 6*(2), 119–134.

Lupton, D. (2018a). Cooking, eating, uploading: digital food cultures. In K. LeBesco & P. Naccarato (Eds.), *The Handbook of Food and Popular Culture* (pp. 66–79). London: Bloomsbury.

Lupton, D. (2018b). 'I just want it to be done, done, done!' Food tracking apps, affects, and agential capacities. *Multimodal Technologies and Interaction, 2*(2). Retrieved from www.mdpi.com/2414–4088/2/2/29/htm

Massey, D. (2005). *For Space.* London: Sage.

Norwood & Brixton Foodbank. (2018, April 23). *Latest figures show 22% increase in foodbank usage.* https://norwoodbrixton.foodbank.org.uk/2018/04/23/latest-figures-show-22-increase-in-foodbank-usage/

Phillipov, M. (2019). Thinking with media: Margins, mainstreams and the media politics of food. In M. Phillipov & K. Kirkwood (Eds.), *Alternative Food Politics: From the Margins to the Mainstream* (pp. 1–20). London: Routledge.

Pilcher, J. (2012). *Planet Taco: A Global History of Mexican Food.* Oxford: Oxford University Press.

Portas, M. (2011). *The Portas Review: An independent review into the future of our high streets.* Retrieved from https://assets.publishing.service.gov.uk/government/uploads/system/uploads/attachment_data/file/6292/2081646.pdf

Rousseau, S. (2012). *Food and Social Media: Your Are What You Tweet.* Lanham: Rowman and Littlefield.

Schneider, T. (2019). Promising sustainable foods: Entrepreneurial visions of sustainable food futures In M. Phillipov & K. Kirkwood (Eds.), *Alternative Food Politics: From the Margins to the Mainstream* (pp. 75–94). London: Routledge.

Schneider, T., Eli, K., Dolan, C., & Ulijaszek, S. (Eds.). (2018). *Digital Food Activism.* London: Routledge.

Shalet, J. (2010). *Brixvill – an experimental platform for young people.* Retrieved from https://productdoctor.co.uk/2010/09/08/brixvill-an-experimental-platform-for-young-people/

Shipley, R. (2000). The origin and development of vision and visioning in planning. *International Planning Studies, 5*(2), 225–236.

Slocum, R. (2007). Whiteness, space and alternative food practice. *Geoforum, 38*, 520–533.

Smith, D. (2014, April 17). It's the end of the world as we know it…and he feels fine. *New York Times.* Retrieved from www.nytimes.com/2014/04/20/magazine/its-the-end-of-the-world-as-we-know-it-and-he-feels-fine.html

Soper, K. (2004). Rethinking the 'good life': The consumer as citizen. *Capitalism, Nature, Socialism, 15*(3), 111–116.

Wilder, C. (2010). A fresh face in south London. *New York Times,* 4 August. Retrieved from www.nytimes.com/2010/08/08/travel/08headsup.html

PART 5

Food futures

12

CONNECTED EATING

Servitising the human body through digital food technologies

Suzan Boztepe and Martin Berg

Introduction

The past two decades have witnessed increased entanglements between food-related practices and digital technologies. Recipe databases, restaurant rating sites, food delivery services, food photos on social media, food games and mobile apps that allow people to monitor and manage their food intake have proliferated, creating not only a growing marketplace, but also new ways of engaging with and thinking about food and eating. More recently, a new breed of digital food technologies has emerged, offering novel ways for people to relate to food and eating, and ultimately adjust their eating habits. Using interconnected sensors, photo recognition, machine learning and/or artificial intelligence, digital food technologies supposedly provide user-specific menu planning, so-called smart recommendations based on perceived individual needs, or other ways of guiding people towards *eating right*. What constitutes healthy eating, however, varies and it is a highly controversial subject. What does eating right mean from the perspective of these connected eating technologies? What do these technologies actually do and how do they envision people's interactions with food?

These questions cannot be answered independently from the underlying assumptions, values and agendas of technology developers and designers (Berg, Fors, & Eriksson, 2016; Lupton, 2014; Purpura, Schwanda, Williams, Stubler, & Sengers, 2011; Schüll, 2012). Designers, by definition, 'consider possible futures, worlds that can be imagined' (Krippendorff, 2006, p. 28) and build them into devices and services using form, affordances, symbols and metaphors, as material expressions of visions and values (Lupton, 2014). Their often-tacit agendas and visions can potentially reshape the 'food landscape' and change how we imagine the eating and living body (Miles & Smith, 2015, p. 119).

By critically analysing three digital food technologies, this chapter establishes an understanding of how the companies behind these technologies, as part of their business model, imagine users and their bodies, the meaning and function of food and the practice of eating. We chose two devices and one service that represent different types of data these technologies use to generate eating advice. These include test data (Habit), food recognition (SmartPlate) and eating behaviour (HAPIfork). They also represent different technologies, such as sensors, photo recognition, augmented reality, machine learning and artificial intelligence.

Digital food technologies are often closed technical systems. In this study, we use critical content analysis (Krippendorff, 2004) to understand how these technologies are sociocultural products underpinned by their creators' ideas and assumptions about their use and value as well as existing agendas and discourses (Berg, 2017; Lupton, 2014). We do this by analysing product descriptions, user guides, sales pitches and various promotional materials, such as blog posts and video clips, through which the products are marketed and their functionalities explained. Specifically, we focus on the ways these technologies are intended to be instilled into people's lives. By doing this, we explicate how the creators of these technologies view the human body and what kind of interaction they envision between people and food. Engaging with these interrelated dimensions of the marketing rhetoric unveils not only the basis for specific business tactics employed to attract and keep customers but also casts light on the business logic of digital food technology developers and how that logic favours certain ways of eating.

From Quantified Self to quantified food

A growing number of digital food technologies claim to revolutionise food systems and eating practices and thus provide personalised food and diet options (Trice, 2016). Much like the burgeoning field of self-tracking and life-logging (see Ajana, 2018b; Lupton, 2016a), digital food technologies are presented as a means for users to navigate through the varying temporalities and contingencies of the everyday life. They are often assumed to solve a series of problems for users in order to help them reach an increased level of wellbeing, and enjoy the alleged benefits of a healthier, more productive and sometimes even optimised life (see Ruckenstein, 2014; Ruckenstein & Pantzar, 2017). Despite growing interest in the meaning and use of digital food technologies, research in this area is still in its infancy, focusing on either technical issues related to their development or social practices and ideals, analysed through a critical sociocultural perspective (see Didžiokaitė, Saukko, & Greiffenhagen, 2018; Lupton, 2018).

In this chapter, we engage with the situated contexts of digital food technologies by asking what they want from their users and how designers and marketers of these devices envision people as they go about their everyday lives, planning their meals and indeed also eating them. Introducing technology into food-related practices means entering a territory that is loaded with emotions, identities, rituals, social relationships, traditions, cultural values and meanings (de Certeau, Giard, & Mayol, 1998; Counihan & van Esterik, 2012). Studies that have taken a critical sociocultural perspective on food technologies, even though limited in number, have begun to paint a picture of how complex sociocultural and biographical factors come into play in human–food–technology interactions (e.g., Lupton, 2018).

The emergence of digital food technologies can be understood as situated in a metric culture (Ajana, 2018a) that is increasingly shaped by numbers, not least through processes of datafication that aim to render nearly all aspects of our lives into data (Beer, 2016; Mayer-Schönberger & Cukier, 2013; Ruckenstein & Schüll, 2017). For example, according to Ruckenstein and Pantzar (2017), people are often assumed to need access to data in order to be able to fully engage in self-reflective practices, that is, understand and reflect on their own bodies and daily actions. Some studies of self-tracking technologies have shed light on how devices and apps of this kind often build on set ideas about people and their everyday lives (Berg, 2017; Lupton, 2014; Fors, Berg, & Pink, 2016; Ruckenstein, 2015; Schüll, 2016). It is not unusual that they envision the user as a nudge-able subject who is assumed to trust her devices not only 'to remotely *track* her but also to keep her *on track*, interrupting the flow of her experience to prompt her – when an algorithmic analysis of her own real-time data deems it necessary – to eat, drink, or rest' (Schüll, 2016, p. 330, original emphasis). Digital food technologies are believed to support new practices and form new perceptions and even shape new models for living (Schüll, 2016), not least by allowing people to engage in self-reflective practices using seemingly clean and neutral numbers as points of reference (Lupton, 2013a, 2013b).

Efforts to develop digital food technologies can be seen as yet another case of technological solutionism (Morozov, 2013), or as an illuminating example of the general techno-scientific ideology. That is because they are based on the belief that every aspect of an 'over-examined life', as Ajana (2018a, p. 2) puts it, can be designed without proper involvement of health professionals (Nikolaou & Lean, 2017; Flaherty, McCarthy, Collins, & McAuliffe, 2018). These ideas are grounded in and afforded by the business models of the companies that design and market technologies of this kind. This necessitates research on food technologies that combines critical sociocultural perspective with analyses of the business models and practices of the companies.

The business logic of digital food technologies seeks to exploit the so-called biocapital (Rajan, 2006) by combining food and personal data in novel ways in the digital knowledge economy (Lupton, 2016b). There are two dynamics at work for value creation in this economy (van Dijck & Poell, 2013, 2016). The first one is *datafication*, which refers to the process of converting every aspect of life into data points (Mayer-Schönberger & Cukier, 2013). The second one is the *deployment of business models* that commodify and convert these data points into capital. While datafication is enabled by technologies that capture human be-haviour and convert it to a quantifiable form, commodification happens through the development of data services (van Dijck, Poell, & de Waal, 2018). Anything, even physical entities, can lead to the development of such data services.

The concept of servitisation (Vandermerwe & Rada, 1988), originally used to describe the transformation of manufactured products into service systems, means that physical products can be turned into services when they are tracked, monitored and cared for. Digital technologies are key enablers of the servitisation as they offer means for connecting and monitoring products (Kryvinska, Kaczorl, Strauss, & Greguš, 2014). A famous example is Rolls-Royce, which moved from selling aero engines to delivering power by the hour and engine care services to its customers thanks to remote tracking. Just as digital technologies allow an engine to be remotely tracked and serviced whenever needed, they now promise the possibility to track the body and its nutritional intake to facilitate an optimised bodily health.

Such service-centred strategies seek to engage people in prolonged acts of consuming through online platforms and apps connected to things (Thrift, 2006). This, in turn, leads to value creation from long-term customers, user valuation and continuous data flow that could be sold to other companies (van Dijck et al., 2018). Businesses of this kind offer their services in exchange for personal information that 'they can endlessly repurpose and combine with other data collections' (van Dijck et al., 2018, p. 110) to create new services.

This servitisation-based business model raises questions about how, and, to what extent, food, eating and the human body can (and should) be quantified, connected and datafied. After all, the practices of cooking and eating are contextually bound and affected by the temporality and sociality of everyday life (Boztepe, 2007; Comber, Hoonhout, Van Halteren, Moynihan, & Olivier, 2013). It has also been shown that engagement with digital food technologies can lead to new calculative practices of food consumption (Niva, 2017). However, the particular ways in which services are envisioned, designed and promoted in light of certain business interests are yet to be explored. This chapter takes a step in this direction by analysing how food technology companies project certain eating practices through their marketing materials.

Digital food technologies

In our attempt to bring to light the future that creators of food technologies envision for us, we will examine three companies. We grouped the digital food technologies currently available on the market in three categories based on the type of data they use for generating eating recommendations, and selected one from each category: Habit, a nutrigenomic company that provides eating advice based on one's DNA, blood tests and lifestyle; SmartPlate, a tableware which aims to help people control, monitor and track the type and amount of food consumed; and HAPIfork, which is another tableware device for monitoring the speed and amount of food consumption. Through these cases we analyse how *connected* eating practices are projected through specific images and discourses of the marketing materials.

Habit – setting the table with DNA

The success of the Human Genome Project and the development of technology for efficiently reading DNA code on a mass scale paved the way for direct-to-consumer (DTC) genomics. The idea that genetic information has massive potential for drug discovery and new treatments made DNA a bio-capital of enormous value (Rajan, 2006). Following this, the DTC genomics industry emerged in the 2000s, promising to improve and individualise healthcare by leveraging knowledge about a person's DNA. The public expectation was kept high by presenting DNA as a holder of the 'secret of life' that would change how 'we live, heal, eat and imagine the future' (Gibbs, 2003, p. 42), despite concerns over its clinical value and ethical implications (see Dickenson, 2013; Ducournau, Gourraud, Rial-Sebba, Cambon-Thomsen, & Bulle, 2013). Companies offering DNA testing to reveal a person's ancestry or to uncover predispositions to certain diseases were soon followed by those telling their customers how to eat, sleep and exercise.

Habit promises to provide its customers with personalised food and nutritional advice based on their DNA, blood tests and lifestyle. After ordering Habit, customers receive a home test-kit, which includes lancets for blood samples, mouth swabs for DNA samples, a tape measure for body dimensions and so on. They also receive a drink called metabolic challenge that includes fats, sugars and carbs to see how well their bodies handle the nutrients. Once they send samples, measures and a survey back to Habit, the data are analysed to create a food profile and eating recommendations. Customers access their profiles and food and recipe recommendations through a digital dashboard on their phones or computers. The dashboard is designed as a coach who identifies the problem, sets a goal, explains the steps, and guides through daily decision-making. For an additional fee, customers also receive advice from a real person. In certain regions, they are also able to order ready-to-eat meals, supposedly tailored to their test results.

Habit's value proposition rests on the assumption that eating is a problematic act involving many uncertainties which should be cast out by a scientific and personalised approach. 'Take the guesswork out of eating right' says the welcome text on the company website (Habit, 2019). That is, the mundane choice of what to eat cannot be left to one's own judgement, because it is believed that there is an accurate way of eating for everyone, defined by what is perceived of as *real* science. The intended customers are those who aspire to do so, but who lack the skills or tools to do it. They care about what they eat, but, as a woman explains in a testimonial video, it is difficult to know what to eat. She concludes that Habit gives her 'the confidence of making the right choices' (Habit, 2019).

The solution to the presented challenge is 'personalized food', because, according to the creators of Habit, eating is so unique that no standard recommendations could meet the complexity of an individual's needs. The testimonial video claims that common-sense healthy eating fails by giving an example of how Habit said carbs are good for her, while healthy eating diets typically advise lowering carbohydrate intake. The CEO's health epiphany is used as further evidence:

> [W]hen he looked in the mirror, he saw a middle-aged white dude with a dad bod that was threatening not just his self-esteem, but his longevity. After DNA and blood tests, he discovered that the best way back to a healthier, energized version of himself started with understanding his body's fundamental needs – and eating the types of foods and nutrients that his body asked for.

These claims suggest that the right way of eating can only be achieved by learning about one's own body. This, in turn, is possible only through quantification of the bodily signals. The body, from this point of view, is seen as a vessel of information: 'Your body knows what it needs' (Habit, 2019). Yet a person is unable to access and decode her own data flow (Viseu & Suchman, 2010). These are locked in the blackbox of one's DNA and metabolism. But with 'science-based' tests 'you can [know] too' (Habit, 2019). The resulting bodily data, however, can be made intelligible only after being connected to nutritional data by a digital platform. There is no mention about the computational technology used for deriving eating recommendations from one's test data. It suffices to state that it uses a 'scientifically validated approach' (Habit, 2019).

The words *science* and *evidence* are used extensively throughout Habit's marketing materials to create a sense of trust, credibility and objectivity. Company scientists are presented as being 'on mission' to transform customers' health (Habit 2019). The ultimate purpose, the website states, is to 'optimize your health with nutrition' (Habit 2019). The underlying metaphor here is that of a person as a machine, the performance and effectiveness of which could be enhanced and perfected. The same metaphor is used in images and videos throughout the website. For example, a video shows the image of a woman talking while callouts – used to connect image and information in technical drawings – pop up in the background, showing properties of her nutritional profile.

Self-discovery and guidance for optimising the body is provided through Habit's digital dashboard. This is where elements of an individual's genetic and blood test data are selected and presented in ways that are accessible and consumable by users. Test results, for example, are presented in numerical form but at the same time explained in everyday language, just as a medical professional would do. Daily calories, weight goal and the nutrient ratio one needs are shown as simple charts to provide a first glimpse into what matters. Habit uses various forms of materialisation that supposedly stimulate users at affective level in order to motivate them to act. One of these is the so-called unique profile. *You are a plant seeker* says the dashboard, creating sense of curiosity and promoting self-discovery. In fact, the user is placed in one of the seven standard diet types created by Habit.

Using the compelling force of imagery, vivid and attractive photos are used to display how an actual plate should look like, or what kind of foods one should eat.

Users also have their own designated so-called hero foods, which are healthy foods supposedly specific to their personal diet type. Note the use of the hero metaphor, which imbues food with powers to save a person: *Food is saviour.* Only the company can show the customers the way to their own liberation. For even better results, users are encouraged to stay connected with the service by syncing it with their Fitbit armbands, using the Habit tracking app snapping photos of everything they eat, and eventually ordering ready-to-eat meals from the company.

SmartPlate and HAPIfork – tableware for regulated eating

In recent years, several efforts have been made to design digital food technologies that aim to make profits by pushing the eating habits to a supposedly healthier but at the same time more controlled direction, such as an algorithmic photo-based dietary tracking system (Ming, Chen, Cao, Forde, Ngo, & Chua 2018) or a wrist-worn device that detects hand-to-mouth eating gestures (Ye, Chen, Gao, Wang, & Cao, 2016). A typical example of this kind is marketed as 'the world's first Intelligent Nutrition Platform that uses advanced photo recognition and AI technology to identify, analyze, and track everything you eat in mere seconds' (Smartplate, 2018) and is called *Smartplate Topview.* The device consists of a three-piece portable plate and countertop dock that pairs with an associated smartphone app. The device uses the smartphone camera to identify up to three meals and/or food stuffs on the plate and allows their nutritional values to be added to a personal dietary journal while also providing personalised guidance on how to move closer to the user's personal health goals.

CEO Anthony Ortiz explains how the device should be used using three kinds of preserved meat as an example. He starts by putting the stationary base on the kitchen counter and says that he wants to show how 'awesome' the Smartplate's image recognition is:

> Notice that these three kinds of meats come from the same animal species. We have deli ham slices, we have salami, and we have prosciutto here. Alright, they look similar in texture, they're red, if you touch them they also feel very similar in the texture. So, how's it possible that smart plate can recognise these different things? So, here we go.
>
> *(Ortiz, 2016a)*

He touches the meat and puts the meats on different tray partitions, sets the weight to zero on the stationary base and then places the tray on top of it. He continues to explain the procedure by taking a picture of the food with the app where the viewfinder matches the contours of the tray. 'Now, within just a few seconds our system is analysing those images, it's sending it to the cloud for recognition', he says, and shows how the app has made a match (Ortiz, 2016a).

The SmartPlate is designed to function like a personal nutritionist that supposedly allows users to 'say goodbye to manual data entry' when measuring calorie intake and to achieve certain health goals. The device and the app are claimed to allow users to 'analyse and adjust meals accurately, easily and quickly' to follow one of the many built-in 'doctor approved dietary programs' or to engage in weight management (Smartplate, 2018). The device can allegedly accurately detect and analyse thousands of foods and food items, including more than 600 national chain restaurant menus in the United States. The need for accuracy is emphasised with reference to the CEO's own family. He explained:

> A few years ago, I had to watch my dad suffer through triple bypass surgery. And that was one of the toughest times for me and my family, but to later learn that his condition was 100% preventable really infuriated me, and I decided to do something about it. Every day you're faced with over two hundred food decisions that have a dramatic impact on the state of your health. And maybe you're one of the hundred million Americans that uses a manual calorie counting app to help you with these food choices. But the problem is these solutions are time-consuming, they're frustrating, and worst of all, they're inaccurate.
>
> *(Ortiz, 2016b)*

The SmartPlate is presented as a device that allows users to control their calorie intake throughout the day without having to spend time on manually calculating the energy values of their food. Instead of manual calculations, the company prescribes nearly automatic, controlled food intake. In this way, the number of calculations one supposedly needs to make in order to eat healthy is reduced to taking a photo of food in the plate, 'And bam! Get all your nutrition information in a snap' (Ortiz, 2016b). With this device, the company not only claims to bring convenience by automation, but also what they call a new category in healthcare: 'culinary medicine'. This idea stretches beyond potential personal health benefits since the company believes that their products can 'make a huge social impact' and 'help drive the change that millions throughout the world so desperately need' (Ortiz, 2016b) in times when chronic and degenerative diseases are on the rise.

The HAPIfork focuses on the mechanics of eating, rather than the actual nutritional content of food, and offers to haptically automate the daily labour of eating (Schüll, 2019). This device, a connected fork, is part of an ecosystem of digital devices and applications that, taken together, claims to help people achieve happiness, wellness and better health. Drawing on the company's interpretation of neurobiological science, the device is marketed as an electronic fork that records eating schedules, monitors eating habits and alerts users through vibrations if they are eating too fast. The device not only nudges users but also tracks eating behaviour and provides a dashboard where each meal can be reviewed. The data tracked and processed include the exact start and finish time of a meal, meal duration, the number of fork servings and the interval in between each fork serving. Based on these data, the device presents information on the so-called success rate and overspeed ratio that are supposed to explain the number of fork servings that have been above the device's recommended threshold. It also provides advice on the interval that users should instead aim for (Hapilabs, 2013). The dashboard consists of diagrams and visual cues that turns eating into an analysable, controlled activity.

Cutlery is one of the few objects that has attracted extensive, high-profile design attention throughout its history (see Dormer, 1993) possibly due to its strong association with impression management. The HAPIfork, however, has turned its back to the search for a refined design language by borrowing its visual language from electronics. Not only does it introduce a new sense of aesthetics similar to SmartPlate, but it also turns attention towards new values of eating that never existed before. For example, the temporality of eating is the focus. The device is designed to help people eat slower and thus, find a pace of life that improves their own and others' overall health and wellbeing by making them 'aware of every good moment' (Hapi, 2018).

Culinary futures

The smart tableware and the nutrigenomic service described above promise better health, weight loss and an energised body by supposedly regulating eating. Despite differences in their approaches, these digital food technologies share a common understanding about their potential customers as aspirants, their bodies as optimisable and eating practices as controllable, as well as an idea about the meaning and functions of food. The products are often marketed and explained through personal experiences of their inventors. The inventors' health epiphanies and personal journeys for a better health are used as the rationale behind their innovations and even for claims of creating a healthier society – one member at a time. In this way, they exemplify what van Dijck and Poell (2016) termed as the 'double-edged logic' (p. 1) of health platforms, meaning that they create an illusion of serving a public interest along with the business one.

The customers portrayed are people who, like the inventors, strive to improve their health and wellbeing through healthy eating. This view of customers as *aspirants* creates the primary condition for exploiting the benefits of the contemporary transformation economy, where value is created through highly personalised offerings that are believed to change the individual for better (Pine & Gilmore, 1999). According to this view, tapping into people's desire to change can result in long-term, high-value economic returns. The suggested potential is in the creation of a sustained consumer market where 'the customer is the product' essentially saying 'change me' (p. 172). In this case, it means changing the body into a healthier, slimmer and more energised one by eating the right food in the right amount at the right pace. The food technologies then act as aspirational forces promising better futures (Schüll, 2019).

Habit, SmartPlate and HAPIfork adopt this logic by portraying the body as a machine that could be optimised and fine-tuned through healthy eating, as prescribed by them. This idea is not new. It has long been promoted by dietitians and public health authorities, creating what Scrinis (2008) termed a nutritionist paradigm. At the heart of this paradigm lies the idea of reducing the body to certain biomarkers and matching those biomarkers with nutrients affecting them. Decades-long efforts to break down food into quantifiable nutrients have made such matching possible. The change, then, depends on the ability to understand and manipulate the relationship between nutrients and biomarkers.

Weight-loss services such as Weight Watchers have used regular monitoring of people's bodies and food intake as a strategy for a long time. The new technologies take this approach one step forward by digitising it. From this perspective, the body must constantly be serviced by food and food data to facilitate change. That is, the body itself becomes the subject of servitisation. The information needed for service already lives in the body though it is in a form that is not directly accessible to consumers (Viseu & Suchman, 2010). Therefore, people are left in doubt about what to eat and they cannot trust their own senses. Digital food technologies then bet on the same insecurity as self-tracking technologies about how to navigate the mundane everyday choices (see Schüll, 2016). This forms the ground for datafication and intervention. The idea that constant and accurate data flow about what is ingested is required for change to take place is at the core of these technologies. Habit solicits data from customers on the grounds that genetic profiling and other tests are the best way to uncover their nutritional needs. The smart tableware devices that we presented earlier depart from the idea that nutritional data needs to be collected and monitored in real time in order for people to adopt healthy eating habits. Eating is ultimately transformed into a data service through a series of calculations and predictions that are believed to be accurate and highly personalised.

Central to the value proposition of the digital food technologies is the claim of high accuracy. SmartPlate, for example, promises an accuracy within the range of 0.5 grams, whereas Habit claims to provide the exact percentage of nutritional distribution on plate. In reality, however, accuracy is not as straightforward as it is promoted. Food items come in various sizes, forms and weights, with ingredients from a variety of sources with different nutritional values depending on season, farming practice and preparation. Things get even more complicated when meals with ingredients from sources that go beyond the standardised menus of restaurant chains and fast food outlets (which, by the way, would get less than excellent scores in terms of their health benefits). On the other hand, since food must be recognisable by the technologies in question, they must be standardised and measurable. Users are thus encouraged to follow the recipes provided by the services, eat at approved restaurants, or choose food items that may be scanned. As a result, individual variations, creative cooking or consuming unrecognisable foods carry the danger of disturbing regulated eating, and, in the long run, putting the servitised body out of service.

The products discussed in this chapter propose that there is a scientifically defined accurate way of eating, in terms of ingredients, quantity, time and even speed. This accurate way of eating can be identified using body signals or data and is measurable. This is not only a marketing strategy but also a way of objectifying the practice of eating as a whole. Stripped from cultural, social and emotional aspects, eating is projected as a purely functional act for optimising the machine-like body. This view stands in stark contrast to sociological understanding of eating as a practice through which one makes 'concrete one of the specific modes of relation between a person and the world, thus forming one of the fundamental landmarks in space-time' (de Certeau et al., 1998, p. 183). When a dinner table (if there is one) is set for the servitised body, the convention followed is not one rooted in culture, sensory experiences or interpersonal relations. Instead, it is a setup that follows and cherishes eating as an isolated experience and food as technologically recognisable, standardised and something for which the quantifiable bodily craves. It is a dinner table shared not by people connected by food, but by standardised food items connected to people through technologies to monitor their every bite.

Conclusion

In this chapter, we looked at three digital food technology companies that supposedly promote healthy eating. Through a critical analysis of their promotional materials, we examined the future they envision for the way to healthy eating. As sociocultural products, we explained how these materials are representative of the underlying assumptions, meanings and concepts of their creators. Our brief analysis, however, has barely scratched the surface in illustrating how the burgeoning new market of technologies of connected eating is exerting new meanings and ways of eating. Further research is needed to understand how such technologies influence and are influenced by our eating behaviour and what this means for how society is organised. Studies on how ordinary people adopt, engage with and incorporate these technologies into their daily lives are particularly needed. Adaptations of or resistance to such technologies may shed light on the dynamics involved in their appropriation. We know little about the assumptions and values of the designers and developers of these technologies. Specifically, design and business practices including vested interests of developers deserves more attention. This may include, but is not limited to, generation, circulation and commercialisation of user data, business model analysis, positioning food technologies as part of platform ecosystems as well as how these technologies are related to political discourse. Personal data security and privacy are also deeply intertwined with these technologies and merit attention. Overall, as digital technology makes inroads on the landscape of food production, provision and consumption, more research is needed to bring to light what our digital food future holds for us.

Further reading

Schüll, N.D. (2018). LoseIt! Calorie tracking and the discipline of consumption. In J. Morris & S. Murray (Eds.), *Appified: Mundane Software and the Rise of the Apps* (pp. 103–114). Ann Arbor: University of Michigan Press.

References

Ajana, B. (Ed.). (2018a). *Metric Culture: Ontologies of Self-Tracking Practices*. London: Emerald.

Ajana, B. (Ed.). (2018b). *Self-Tracking: Empirical and Philosophical Investigations*. London: Palgrave Macmillan.

Beer, D. (2016). *Metric Power*. London: Palgrave Macmillan.

Berg, M. (2017). Making sense with sensors: Self-tracking and the temporalities of well-being. *Digital Health*, *3*, 1–11.

Berg, M., Fors, V., & Eriksson, J. (2016). Cooking for perfection: Transhumanism and the mysteries of kitchen mastery. *Confero*, *4*(2), 111–135.

Boztepe, S. (2007). Toward a framework of product development for global markets: A user-value-based approach. *Design Studies*, *28*, 513–533.

Comber, R., Hoonhout, J., Van Halteren, A., Moynihan, P., & Olivier, P. (2013). Food practices as situated action: Exploring and designing for everyday food practices with households. *Proceedings of the SIGCHI Conference on Human Factors in Computing Systems*. ACM, pp. 2457–2466.

Counihan, C., & van Esterik, P. (Eds.). (2012). *Food and Culture: A Reader*. London: Routledge.

de Certeau M., Giard, L., & Mayol, P. (1998). *The Practice of Everyday Life, Vol. 2., Living and Cooking* (trans J.T. Tomasik). Minneapolis: University of Minnesota Press.

Dickenson, D. (2013). *Me Medicine vs. We Medicine: Reclaiming Biotechnology for the Common Good*. New York: Columbia University Press.

Didžiokaitė, G., Saukko, P., & Greiffenhagen, C. (2018). Doing calories: The practices of dieting using calorie counting app MyFitnessPal. In B. Ajana (Ed.), *Metric Culture: Ontologies of Self-Tracking Practices* (pp. 137–155). Bingley, UK: Emerald Publishing.

Dormer, P. (1993). *Design since 1945*. London: Thames & Hudson.

Ducournau, P., Gourraud, P.-A., Rial-Sebbag, F., Cambon-Thomsen, A., & Bulle, A. (2013). Direct-to-consumer health genetic testing services: What commercial strategies for which socio-ethical issues? *Health Sociology Review*, *22*, 75–87.

Flaherty, S.J., McCarthy, M., Collins, A., & McAuliffe, F. (2018). Can existing mobile apps support healthier food purchasing behaviour? Content analysis of nutrition content, behaviour change theory and user quality integration. *Public Health Nutrition*, *21*(2), 288–298.

Fors, V., Berg, M., & Pink, S. (2016). Capturing the ordinary: Imagining the user in designing automatic photographic lifelogging technologies. In S. Selke (Ed.), *Lifelogging* (pp. 111–128). Wiesbaden: Springer.

Gibbs, N. (2003). The DNA revolution: The secret of life. *Time*, *16*(7), 42–45.

Habit. (2019). H*ABIT: Food personalized*. Retrieved from www.habit.com/

Hapi. (2018). *HAPI.com: It's time to join the HAPI revolution*. Retrieved from www.hapi.com/

Hapilabs. (2013, December 12). *HAPIfork FAQ: What does the HAPIfork track*. [Video file]. Retrieved from https://youtu.be/v-XOwCra3fs

Krippendorff, K. (2004). *Content Analysis: An Introduction to Its Methodology*. Thousand Oaks, CA: Sage.

Krippendorff, K. (2006). *The Semantic Turn: A New Foundation for dEsign*. New York: CRC Press.

Kryvinska, N., Kaczorl, S., Strauss, C., & Greguš, M. (2014). Servitization: Its rise through information and communication technologies. In M. Snene & M. Leonard (Eds.), *Exploring Services Science* (pp. 72–82). New York: Springer.

Lupton, D. (2013a). Quantifying the body: Monitoring and measuring health in the age of mhealth technologies. *Critical Public Health, 23*, 393–403.

Lupton, D. (2013b). The digitally engaged patient: Self-monitoring and self-care in the digital health era. *Social Theory & Health, 11*(3), 256–270.

Lupton, D. (2014). Apps as artefacts: Towards a critical perspective on mobile health and medical apps. *Societies, 4*(4), 606–622.

Lupton, D. (2016a). *The Quantified Self: A Sociology of Self-Tracking.* Cambridge: Polity.

Lupton, D. (2016b). The diverse domains of quantified selves: Self-tracking modes and dataveillance. *Economy and Society, 45*(1), 101–122.

Lupton, D. (2018). 'I just want to be done, done, done!' Food tracking apps, affects and agential capacities. *Multimodal Technologies and Interaction, 2*(2). Retrieved from www.mdpi.com/2414–4088/2/2/29/

Mayer-Schönberger, V., & Cukier, K. (2013). *Big Data: A Revolution that will Transform How We Live, Work, and Think.* Boston: Houghton Mifflin Harcourt.

Miles, C., & Smith, N. (2015). What grows in Silicon Valley? In H.L. Davis, K. Pilgrim, & M. Sinha. (Eds.), *The Ecopolitics of Consumption: The Food Trade* (pp. 119–138). London: Lexington Books.

Ming, Z.Y., Chen, J., Cao, Y., Forde, C., Ngo, C.W., & Chua, T.S. (2018). Food photo recognition for dietary tracking: System and experiment. In K. Schoeffmann et al. (Eds.), *MultiMedia Modeling. MMM 2018. Lecture Notes in Computer Science* (pp. 129–141), vol. 10705. Cham: Springer.

Morozov, E. (2013). *To Save Everything Click Here: Technology, Solutionism and the Urge to Fix Problems that Don't Exist.* London: Penguin Books.

Nikolaou, C.K., & Lean, M.E. (2017). Mobile applications for obesity and weight management: Current market characteristics. *International Journal of Obesity, 41*(1), 200–202.

Niva, M. (2017). Online weight-loss services and a calculative practice of slimming. *Health, 21*, 409–424.

Ortiz, A. (2016a, July 4). *SmartPlate deli meat demo.* [Video file]. Retrieved from https://youtu.be/zozLHEf77jc

Ortiz, A. (2016b, November 16). *SmartPlate topview.* [Video file]. Retrieved from https://youtu.be/76NLQwB5S5A

Pine, B.J., & Gilmore, J.H. (1999). *The Experience Economy.* Cambridge, MA: Harvard University Press.

Purpura, S., Schwanda, V., Williams, K., Stubler, W., & Sengers, P. (2011). Fit4life: The design of a persuasive technology promoting healthy behavior and ideal weight. *Proceedings of the SIGCHI Conference on Human Factors in Computing Systems* (pp. 423–432). New York: ACM.

Rajan, K.S. (2006). *Biocapital: The Constitution of Postgenomic Life.* Durham: Duke University Press.

Ruckenstein, M. (2014). Visualized and interacted life: Personal analytics and engagements with data doubles. *Societies, 4*(1), 68–84.

Ruckenstein, M. (2015). Uncovering everyday rhythms and patterns: Food tracking and new forms of visibility and temporality in health care. *Studies in Health Technology and Informatics, 215*, 28–40.

Ruckenstein, M., & Pantzar, M. (2017). Beyond the quantified self: Thematic exploration of a dataistic paradigm. *New Media & Society, 19*, 401–418.

Ruckenstein, M., & Schüll, N.D. (2017). The datafication of health. *Annual Review of Anthropology, 46*, 261–278.

Schüll, N.D. (2012). *Addiction by Design: Machine Gambling in Las Vegas.* Princeton, NJ: Princeton University Press.

Schüll, N.D. (2016). Data for life: Wearable technology and the design of self-care. *Bio Societies, 11*(3), 317–333.

Schüll, N.D. (2019). HAPIfork and the haptic turn in wearable technology. In M. Boucher, S. Helmreich, L. Kinney, S. Tibbits, R. Uchill, & E. Ziporyn (Eds.), *Being Material* (pp. 70–75). Cambridge, MA: MIT Press.

Scrinis, G. (2008). On the ideology of nutritionism. *Gastronomica, 8,* 39–48.

Smartplate. (2018). *Smart plate.* Retrieved from https://getsmartplate.com/

Thrift, N. (2006). Re-inventing invention: New tendencies in capitalist commodification. *Economy and Society, 35*(2), 279–306.

Trice, R. (2016). Why we need an internet of food. Retreived from www.forbes.com/sites/themixingbowl/2016/10/14/why-we-need-an-internet-of-food/#2017b03a44b1

Vandermerwe, S., & Rada, J. (1988). Servitization of business: Adding value by adding services. *European Management Journal, 6*(4), 315–324.

Van Dijck, J., & Poell, T. (2013). Understanding social media logic. *Media and Communication, 1*(1), 2–14.

Van Dijck, J., & Poell, T. (2016). Understanding the premises and promises of online health platforms. *Big Data & Society, 3*(1), 1–11.

Van Dijck, J., Poell, T., & de Waal, M. (2018). *The Platform Society: Public Values in Connective World.* New York: Oxford University Press.

Viseu, A., & Suchman, L. (2010). Wearable augmentations: Imaginaries of the informed body. In J. Edwards, P. Harvey, & P. Wade (Eds.), *Technologized Images, Technologized Bodies* (pp. 161–184). New York: Berghahn Books.

Ye, X., Chen, G., Gao, Y., Wang, H., & Cao, Y. (2016). Assisting food journaling with automatic eating detection. *Proceedings of the 2016 CHI Conference Extended Abstracts on Human Factors in Computing Systems.* ACM, pp. 3255–3262.

13

FROM SILICON VALLEY TO TABLE

Solving food problems by making food disappear

Markéta Dolejšová

Introduction

Digital technologies increasingly contribute to food cultures and practices. Diet tracking apps, smart kitchenware, online 'diet hacking' forums – these are just a few examples from the burgeoning realm of technology products designed to datafy our experiences with food. Pivotal to this food-technology innovation is the Silicon Valley startup sector that has developed major interests in designing solutions for 'better' – sustainable and healthy as well as convenient and playful – food futures. For the Silicon Valley foodpreneurs (food-tech entrepreneurs), food became a hackable object 'poised for massive disruption and change' (Musk, 2016, np). Given the problems with unsustainable food production and malnutrition constraining contemporary food systems (FAO, 2017; WHO, 2017), these proposals are certainly not ill intended. Nevertheless, the idea of solving societal problems around food through the 'right algorithm' (Fumey, Jackson, & Raffard, 2016) has raised many doubts and critiques among food professionals as well as the lay public. These concerns include the potential negative impacts of human–food automation on consumers' responsibility for their food practices as well as on the broader sociocultural frameworks of food production and consumption (Fotopoulou & O'Riordan, 2017; Gioia, 2015; Lupton, 2018b; Miles & Smith, 2015).

Human–food automation is an emerging topic in academic food research, and there has been a growing interest among scholars in debating the opportunities and risks that digital technology presents to contemporary food cultures (Comber, Choi, Hoonjout, & O'Hara, 2014; Dolejšová, Khot, Davis, Ferdous, & Quitmeyer, 2018; Lupton, 2018a). My aim in this chapter is to extend these food-tech reflections with insights from my long-term ethnographic study of the Complete Foods diet – a quantified powdered meal replacement originating from the Silicon Valley realm. Complete Foods (CF) meals are typically delivered as powdered blends of macro and micronutrients (e.g., rice protein instead of rice; oat flour instead of bread; maltodextrin as a source of carbohydrates) to be mixed with water and consumed as liquid shakes. CF products are purported to be nutritionally complete and enable consumers to bypass – fully or partially – the consumption of conventional solid food. Through their powdered form, CF allow for exact measures of ingredient ratios and, hence, for a quantified control over a dieter's nutritional intake. CF meals thus resemble 'nutritional puzzles' where users combine preferred macro and micronutrient ingredients to achieve a diet that best suits their personal needs and wants (e.g., low-carb, ketogenic, vegan).

Silicon Valley disrupting the table

The Silicon Valley startup sector has been on the cusp of food technology innovation since 2012. In 2015, food-tech was announced as one of the fastest-growing technology industries with over $5.7 billion in investments (CB Insights, 2016). This focus on food issues is a continuation of the recent shift of Silicon Valley production beyond purely digital and internet-oriented realms towards broader social issues, such as public health or environmental sustainability (Miles & Smith, 2015). In the specific area of nutrition and diet optimisation, food startups build on the burgeoning trend of health hacking and data-driven self-enhancements that is perhaps best represented by the Quantified Self movement of self-tracking enthusiasts (Lupton, 2014).

Among the many examples of diet-oriented technosolutions are online diet planning platforms such as Lose It, My Plate and MyDietCoach that enable users to monitor their metabolic reactions to consumed food and discuss their results with others. These technologies have been shown to provide a means for consumers to take better care of their health and reach their personal dietary goals (Lupton, 2018b). Similar diet personalisation is enabled by direct-to-consumer (DTC) nutrigenomic services such as DNAFit, Habit or My Gene Diet that provide health and diet recommendations based on users' decoded DNA. Users can learn about their genetic predisposition to digest different nutrients and discuss their results on related online community forums – functions that can help food non-experts to better understand their health and plan their diets (Kuznetsov, Kittur, & Paulos, 2015).

Growing attention in the Silicon Valley food-tech realm has been recently given to the possible impacts of nutrition on human cognition (see activities around Biohacker Summit, Biohackers Collective, Biohack.me). Often discussed under the umbrella term 'self-biohacking', practices in this area include a supplementing of daily meals with nootropics (or so-called 'smart drugs'), adaptogens (fungi and herbs purported to protect the body from stress), or sub-perceptual doses of psychedelics to improve mental focus – a practice known as 'psychedelic microdosing' (Anderson, 2018; Sovijärvi, Arina, & Halmetoja, 2016).

Self-biohacking typically involves a self-experimentation with these supplements and is popular among consumers aiming for self-enhancement *ad optimum*, beyond the standard human health average (Wiesing, 2008).

These food technologies and diet solutions are typically promoted as a means to achieve better control over one's personal health. Through measuring, quantifying and 'hacking' their everyday food practices and metabolic processes, users have the option to develop a – supposedly precise – data-driven understanding of their bodies and diets. Along with this food datafication, users are offered a moral comfort with making the 'right' healthy food choices (Gioia, 2015). Often scaffolded by the techno-capitalist rhetorics of efficiency and libertarian self-care as well as the transhumanist ideals of technological self-perfection, food-tech solutions are further marketed as a way to lower users' dependencies on expert food systems (Andersen, 2018; Rhinehart, 2013; Sovijärvi et al., 2016). As I discuss below in my case study analysis, some consumers see expert food recommendations and healthy eating standards defined by nutrition specialists as too vague and untrustworthy.

Despite these opportunities for a user's health improvement and better information transparency, solutions for quantified diets have been criticised for reducing food into a purely nutritional object and disconnecting users from the broader social and cultural contexts of food consumption (Fotopoulou & O'Riordan, 2017; Miles & Smith, 2015). What happens to our common-sense knowledge of food when we keep outsourcing our mundane food decisions through algorithms? What changes does automation and quantification of food experiences present to social food practices and traditions? In the remainder of this chapter, I will unpack the opportunities and controversies of food-tech innovation on the specific example of the CF diet, which is closely related to the above food-tech practices. CF users often monitor their diets through some self-tracking technology and experiment with various self-enhancement techniques, such as customisation of CF powders according to their nutrigenomics results or supplementing with nootropics.

From Soylent to Complete Foods

The CF diet came into a wider public recognition in 2013, along with the launch of the first CF product called Soylent. The Soylent diet was originally introduced as a self-experiment of software engineer Robert Rhinehart and later turned into a $72 million business backed by Silicon Valley's leading venture capitalists (Crunchbase, 2018). In his blog post titled 'How I stopped eating food', Rhinehart (2013) presented Soylent as a solution for his food frustrations and his inability to prepare and consume meals that would fulfil his healthy eating standards. As a busy tech-worker, he detailed that his daily diet consists mostly of fast food meals and frozen dinners. Given his professional background, he decided to approach his personal diet as an engineering problem and develop an all-your-body-needs nutritional product, using the official United States Department of Agriculture (USDA) diet recommendation as a template. In the blog post, Rhinehart published a report of his month-long diet self-experiment where he consumed nothing but his Soylent concoction. According to him, the diet had positive impacts on his physical and mental fitness (decrease in body weight, better fitness performance and sleep patterns), personal time management (no time spent on cooking, shopping, and eating), financial budget (no spendings on costly 'healthy' foods), as well as on his sustainable food practices (zero food waste, less food packaging).

Rhinehart's Soylent experience caused a wave of enthusiastic reactions on social media and quickly attracted consumers' attention. His idea of replacing daily meals with a quick-fix Soylent potion appealed especially to other startup workers, who saw it as a viable way to optimise their busy lives. The option to use Soylent to quantify their daily food intake further resonates with self-tracking practitioners using digital devices to monitor their daily energy intake and expenditure (details in Dolejšová & Kera, 2017). After a successful crowd-funding campaign and seed funding received from venture capitalists, Rhinehart founded the startup company Rosa Labs to distribute Soylent commercially (Luzar, 2013). Since then, the drinkable powdered-based diet has grown into a worldwide phenomenon that came to a broader recognition as Complete Foods diet (CFS, 2018). To date, there exist over 100 independent startups offering CF products of diverse nutritional compositions and flavours (see www.blendrunner. com for an overview). Embracing the techno-optimism typical for Silicon Valley innovation centres, CF producers have promoted their products as solutions to various food problems and promised to deliver 'better nutrition through innovation' (see the campaign of the Australian-based CF company Aussielent, www. aussielent.com.au).

While the growing marketplace of commercial CF products enabled the convenience of purchasing pre-mixed powders and ready-to-drink CF meals, many consumers still prefer to make their own customised CF formulas. Following the DIY self-experimental approach introduced by Rhinehart, these consumers prepare their CF powders from scratch and share the recipes at a community-driven online platform www.completefoods.co (to date, the site contains over 9,000 recipe entries). Besides the Completefoods site, the CF community (both DIYers and consumers purchasing commercial products) gathers at various discussion forums. Among the most popular – according to a recent CF survey (CFS, 2018) – are the Soylent Discourse (http://discourse.soylent.com/) and the Soylent Reddit (www.reddit.com/r/soylent/). Here, forum users share their CF experiences, discuss general issues related to food and nutrition, and help each other to make sense of their personal diets.

While CF is not a digital food technology product per se, these online sites have been central to the gradual development of the experimental diet (Dolejšová & Kera, 2017). The use of online forums for knowledge sharing related to alternative food practices is not unusual. Forums have discussed vegan diets (Sneijder & Te Molder, 2009), freeganism and dumpster diving (Nguyen,

Chen, & Mukherjee, 2011), as well as food-sharing communities (Ganglbauer, Fitzpatrick, Subasi, & Güldenpfennig, 2014). The CF group places their online forums into the centre of their practice and emphasises the peer-to-peer character of the CF diet. In 2013, Rhinehart offered his original Soylent formula as an 'open source' prototype to be further tweaked and developed by others. The online sharing of personalised user-generated recipes within the newly formed powdered foods community was essential for the further expansion of the CF phenomenon (Dolejšová & Kera, 2017).

Fascinated by the crude pragmatism and asceticism of CF as an edible data-driven solution to everyday food problems, I conducted an ethnographic study aiming to understand the motivations and aspirations embraced by the CF community. Between December 2014 and March 2017, I completed interviews with 65 CF dieters, a qualitative content analysis of the three online forums, three hands-on workshops focused on making of DIY powders, and two autoethnographic self-experiments where I probed the opportunities and limitations of CF on myself. In this chapter, I focus primarily on the data gathered through the interviews and my own CF self-experimentation, with an aim to provide a first-hand account of common CF practices and issues. I discuss how CF practitioners handle health safety and data security risks of their unusual diet, and how they negotiate between data-driven accuracy and gastronomic aspects of food consumption. I show that despite the risks involved, CF offers an opportunity for the dieters to enhance their nutrition literacy and develop confidence with diet personalisation. Drawing on my findings, I suggest that the CF phenomenon, along with its opportunities and drawbacks, aptly illustrates the ambivalent character of Silicon Valley food-tech innovation and its impacts on food cultures.

Case study: Complete Foods

Methodology

As part of my long-term CF ethnography, I conducted interviews with 65 CF consumers (ethics approval was granted by the National University of Singapore's Institutional Review Board). During the interviews, my focus was primarily on participants' experiences with the CF diet and the three community forums. The 65 interviewees were recruited via a convenience sampling at the three online CF forums (through personal messages sent over the forums' messaging system) and a subsequent snowballing. The interviews took place between 2014 and 2017, either face-to-face (n=42) or over Skype (n=23). The live interviews were held in 15 cities across North America (n=20), Europe (n= 42), and Southeast Asia (n=3). The research sample was largely male-skewed and only six participants self-identified as females – a demographic asymmetry also found in the CF survey (CFS, 2018). Respondents were between 20 and 45 years of age and a majority worked or studied in the technology sector. However, there were also artists, journalists and a physiotherapist in the interview sample. Each interview took between 60 and 90 minutes and was audio-recorded and transcribed; transcripts were coded in the CAQDAS software NVivo and analysed using the thematic analysis approach (Braun & Clarke, 2006). Starting with the initial familiarisation with data and generating of initial codes, the analysis process resulted in four main themes that capture the motivations, opportunities and also risks underlying the CF diet (as I discuss below). All participants were given numbers (P1, P2 and so on) to preserve their anonymity.

The autoethnographic part of my fieldwork comprised two month-long sessions of CF diet: during the first self-experiment, I consumed commercial CF products of various brands (Soylent, Mana, Queal, Superbodyfuel, Synectar). For the second period, I consumed a DIY formula that I prepared based on knowledge sourced from the CF forums, especially Completefoods. For both periods, I consumed CF almost exclusively, with occasional 'cheats' where I ate conventional solid food (six times across the two months). Both self-experiments took place in Singapore (1 to 30 August 2015; 21 February to 21 March 2017). During the two months, I maintained a daily diet journal in a simple Google document, where I logged my daily calorie intake (the type and number of CF portions consumed) and expenditure (based on physical activity). I also tracked my body weight, social activities and mood. Alongside these observations, I aimed to develop an insider's perspective on CF. In what follows, I use the notes from my journal to complement my findings from the interviews, which I consider to be the primary data source for my analysis.

Findings and discussion

The case study participants had diverse experiences with CF: 40 interviewees were DIYers making their own customised meals; 25 were purchasing ready-to-drink CF products. Of the 40 DIYers, 18 were selling their powders commercially, either as a small-scale business operated solely by the participant or as part of a larger CF startup company. Among the latter was also the founder of Soylent Robert Rhinehart. The duration of participants' CF diet ranged from one month to more than three years. Only eight participants were consuming a 100 per cent CF diet and used powdered meals as a full food replacement; the rest were using CF as a food supplement: 32 people were replacing one or two meals daily; 15 were occasional users (less than seven CF meals weekly). None of the participants saw CF as a total 'solution': even those eight people claiming to use CF as a full food replacement said that, from time to time, they eat what they crave.

Motivations: escaping food and information overload

Participants had distinct motivations to follow the diet that most often revolved around their desire to improve their personal health and optimise their daily schedules. DIYers valued the option to customise their powders to their personal needs and wants; for customers purchasing commercial CF products, convenience scored higher than customisation. In all cases, participants saw CF as a 'better' alternative to conventional meals:

> I now view every meal in comparison to powdered foods. Every meal is more expensive, less nutritious and more time-consuming to make … the freedom that comes with products like Soylent is indescribable.
>
> *(P12)*

Another motivation shared by participants was the option to use CF to stream-line the variety of diverse and often competing food information that was avail-able to them:

> I feel overwhelmed by the volume of misleading food information. There is such a wide variety of opinions on what is healthy for whom … first you hear that fat is bad, then suddenly it is good, but sugar is bad. And what about proteins – well protein is always good for you, but hey, careful, only some protein! CF offers the way out of this maze: no variety, no need to make a choice.
>
> *(P33)*

Consumers' confusion with official food information such as dietary guidelines and nutrition labels is a frequently discussed topic (e.g., Leek, Szmigin, & Baker; 2015; Spiteri Cornish & Moraes, 2015). As I mentioned earlier, quantified diet solutions are typically designed to help users make sense of available food infor-mation, by suggesting personalised diets tailored to their specific needs. The CF diet goes one step further and reduces the scope of available food options into the uniform food powders:

> Soylent helps me to differentiate between food and nutrition, something that comes together usually. I do not have to 'use food' for nutrition, and I do not have to think about nutrition while having food. Nutrition is the default-stuff and fuel that my body lives on, food is the stuff I enjoy occasionally.
>
> *(P8)*

This utilitarian approach to gastronomy presents a prime example of what Gry-gory Scrinis (2013) calls 'nutritionism' – the reductionist view of food as a sum of nutrients rather than a complex social and cultural object. The extreme form of nutritionism embodied by the CF solution has been a subject of frequent critique. Alice-Azania Jarvis suggests that consuming CF implies 'rejecting gut instinct and, instead, fuelling the gut with data' (2014, np). Christopher Miles and Nancy Smith argue that the CF abstinence from cooking, eating and other sensory food experiences makes the human–food engagement intrinsically 'disembod-ied' (2015, p. 8). The CF model of instant quantified nutrition also implies the abandonment of food as a social and convivial practice: instead of sharing meals with others as a form of social bonding (Barthes, 1997), CF powders are designed for quick consumption, either 'at the desk' or 'on the go'.

In contrast to the above critiques coming from outside of the CF community, participants did not see the CF nutritionism as limiting and constraining their food experiences. Although some of them indeed saw their CF mealtimes as a (desirable) 'solitary event intended to deliver the allotted daily calories' (P46), others suggested that CF helped them to build 'a stronger relationship with the whole idea of food and eating' (P4). For P4, regular consumption of CF meals turned the 'rare moments of eating and preparing normal food into festive and enjoyable social occasions'. Another participant mentioned that drinking CF at work during lunch with colleagues (who eat conventional food) usually serves as a 'great conversation starter that helps to avoid the struggles with finding good small talk topics' (P61). In other words, the abandonment of food consumption as a 'compulsory' everyday task helped some CF users enjoy ordinary food practices and social occasions even more than when eating a 'regular' diet.

I had a similar experience during my two self-experiments: most of my peers were curious to find out more about the content of my CF pitcher during our shared lunchtimes. Drinking my lunches enhanced my social reputation of the 'food experimenter', which I somewhat enjoyed – for me, CF served as a convenient tool to initiate conversations about my research. On the other hand, I deliberately skipped many dinner events and night-outs during my two months of CF. Sipping Soylent and water during casual dining events or in a bar simply felt counterintuitive. The skipped dinners gave me more time to do my work, which I welcomed. On the other hand, I often recalled Jeff Sparrow's commentary on the CF diet as a form of 'Self-Taylorism' (2014, np) during those solitary working nights.

Learning through self-experimentation and peer knowledge sharing

Related to their relief at escaping the 'food information overload', participants often expressed their distrust in the official dietary recommendations issued by food policy authorities. A participant based in Rotterdam mentioned his scepticism of European policy standards: 'The EFSA [European Food Safety Authority] regulations are scientifically outdated and politicized; it's like 5–10 years behind the current research' (P34). While various national dietary guidelines are often used as a baseline for CF recipes, participants (both DIYers and customers) saw such guidelines as too general and in need of further refinement. In other words, these guidelines served as a 'mere' starting point for further self-experimentation:

> I simply trust myself more than anybody else. How my body reacts on a particular diet is more important than what some government health expert or scientist says ... I am quite skeptical of any fixed institutional rules for how to eat healthily.

(P58)

P58's statement illustrates the approach of the CF diet community towards expert food knowledge systems. CF dieters inherently rely on nutrition science and are informed by existing dietary standards; however, rather than trusting expert food authorities fully, they prefer a knowledge derived through self-experimentation that they further corroborate in the online peer community. For the majority of respondents, the CF forums served as a credible information source:

> It's a great source of information – people share Soylent recipes and expe-riences, discuss nutrition studies, help each other with issues ... I don't see any reason why not trust the community. To me, it seems that members there don't have any hidden agendas, unlike food politicians and various experts who are trying to sell you stuff.
>
> *(P29)*

The option to crowdsource nutrition information in the forums provided an opportunity for learning. Participants often mentioned that they did not know much about nutrition before they started with CF. Given the novelty of the diet and the lack of any widely accepted consensus regarding 'good' balanced CF rec-ipes, they welcomed the opportunity to double-check the validity of their (DIY or purchased) powder recipes in the forums. These peer knowledge exchanges made them feel more aware of nutrition principles:

> I needed to read a lot about micronutrients to become confident with my formula. I used mostly Pubmed [an online database of published scientific papers] to find what I need, and the CF forums are a great information source as well. I suppose I am now more aware of how nutrients work in the human diet ... I definitely trust myself more.
>
> *(P19)*

In my CF practice and as a nutrition science amateur, I found the CF forum extremely helpful, especially during the second-month session, for which I de-signed my own DIY formula. To make my personalised blend, I used the Com-pletefoods recipe platform and searched through user-uploaded recipes that were close to my desired diet. I was looking specifically for a vegan recipe with a lower carbohydrates content, and ended up replicating a DIY recipe called Keto Fuel Plain (www.completefoods.co/diy/recipes/keto-fuel-plain), where I replaced fish oil with coconut oil. At the Soylent discourse forum, I checked for further low-carb vegan ideas, which drove me to experimentation with additional in-gredients, such as nori seaweed. While the forums helped me to design a recipe that satisfied my needs and wants, I felt somewhat uneasy relying on nutritional recommendations shared by other forum users – amateur nutrition enthusiasts who developed their knowledge through n=1 self experiments. I was curious how other CF dieters feel about the safety of sourcing dietary knowledge in the forums.

Health safety risks in the community of nutrition enthusiasts

Despite their generally positive attitude towards the forums as a welcoming and supportive space, some participants were quite concerned with the validity of nutrition knowledge shared by the users:

> When I made my first CF formula, I was first getting a quite heavy acidity reflux, and I did not know why … so I asked at the Discourse forum and got a bunch of responses: to replace my oat protein with hemp or rice protein; switch my fibre source from psyllium husk to acacia gum; add antacid supplements like Maalox … I tried those, and I got rid of the reflux after all. But sure, I was also thinking – well, who knows what's gonna happen to me now … (laughing).
>
> *(P56)*

Although none of the interviewees reported any major health issues related to their diet, most of them were aware of the potential health risks involved. Such risks of 'amateur' diet advice are higher in the case of DIYers making their own powders from scratch. However, even the CF products offered for commercial purchase have not yet been subjected to any rigorous health safety testing or clinical trials (this lack of health safety validation was recently reflected in several cases of restricted Soylent distribution: for instance, the Canadian government disallowed further shipments of Soylent in October 2017). Nevertheless, these issues were not perceived as a major problem during my interviews – on the contrary, legal restrictions of CF distribution were seen as unnecessary bureaucracy and only confirmed participants' scepticism concerning expert food authorities. Following their self-experimental approach, CF dieters understood risk as an essential part of producing knowledge about their bodies and diets:

> Sure, it is possible that I have miscalculated the nutrition profile in my Soylent. But how is that worse than eating food from the filthy kebab place around the corner that might give you food poisoning? Experimenting with my powder made me more conscious about food safety in general … the point is that you need to study these issues by yourself, not just buy and eat what is out there … you need to do your research and understand what you eat.
>
> *(P2)*

Such claims about the need for health self-care were often driven by participants' libertarian desires to have a better control over their life: 'Of course there are some health risks, but as long as I am the one concerned and affected, I can deal with those risks. It is my body, after all' (P10). Despite their willingness to take full responsibility for their health, and bear consequences of potential failures in their self-experimentation, none of the participants had a clear idea on what to do in case CF caused them a serious health problem. The majority of responses consisted of rather vague claims such as: 'If I start feeling bad, I will start thinking about that' (P56) or statements that were contradictory to the idea of self-care: 'If I get sick, I guess I will need to go see my GP' (P43). This lack of a clear plan in case things went wrong shows the limited reach that the peer-help provided in CF forums has in a real-life context: should an actual health problem occur, the affected users become invisible to the community, and troubleshooting needs to take place elsewhere – for instance in a professional healthcare facility.

Privacy and security of sharing personal diet tracking data

Besides CF recipes and experiences, forum users often share spreadsheets with their personal diet tracking data (e.g., body weight, sleep quality, mood, energy levels) and medical records (e.g., blood tests or DNA results obtained via nutrigenomics services). A total of 21 participants said that they track their metabolic reactions to the CF diet regularly and share the data via the Soylent Discourse forum (that has a user-friendly interface for such data sharing and offers the option for file uploads). For P19, self-tracking instilled confidence in his diet practice:

> I track my Soylent routine every day, I also track my steps, my jogs, my sleep cycle – the more info I get, the more I feel better about myself ... the Discourse forum provides a great space to look into other people's data, compare your results, and discuss issues.

By sharing their personal data, participants typically aimed to find possible reasons behind the outlier values in their tracking records such as high cholesterol or low blood pressure. Online sharing of personal self-tracking data is a common part of quantified diet practices, as it was discussed in the context of Quantified Self groups (Lupton, 2014) as well as nutrigenomic services (Stewart-Knox et al., 2015). While discussing the possible uses of such data sharing for individual self-discovery, the above authors highlight related privacy risks: the information contained in personal self-tracking data can reveal sensitive details about one's health. Stewart-Knox and colleagues discuss the potential risks of such personal data being misused for commercial gain or falling into the hands of health insurers and government agencies. Health insurance companies might refuse to insure a person who is likely to develop cancer according to her genetic data evidence; employers might withdraw from hiring an employee with mood swings.

Reflecting on these privacy concerns, I was interested in the data security policy of the Soylent Discourse forum. Upon scrutiny, one quickly discovers that the forum does not provide much clarity regarding the security of users' data-sharing activities. It is not clear how and with whom the forum owners (the official Soylent producer Rosa Labs) currently share user data, nor what their future intentions are in this regard. These ambiguous policies were noticed also by the interviewees:

> Open sharing of my personal biodata certainly makes me feel uncomfortable … I hope we will eventually put together a process that would allow people to submit such data anonymously and securely.
>
> *(P22)*

Such concerns are frequently discussed in the forum itself; however, nobody has so far proposed any ideas for how to improve the forum's security parameters. These concerns of CF dieters about the lack of control over their self-tracking data contrast with their desire to have better control over their health. While on the one hand they value the option to achieve a data-driven control over their food practices, on the other hand, the powdered foods dieters paradoxically lose control over their own biological data. Lupton (2018a) discusses a similar situation in the context of food-tracking apps, where users get frustrated with 'the fear of becoming too controlled' via the data-tracking algorithms and archiving functions of their apps.

Conclusion

In this chapter, I have discussed the diverse motivations, ideals, as well as risks underlying the Complete Foods diet. I have shown that CF users deliberately undertake risks and give up some gastronomical pleasures for the sake of having better control over their bodies and diets. My study participants – people with access to information resources and purchasing power – saw CF as a way to escape a burdensome overload of available food products and healthy eating recommendations. While the idea of 'solving food problems by making food disappear' and reducing available food options into a simplified nutritional powder might sound counterintuitive, CF helped participants to become more confident about their diets. Instead of choosing official food and health recommendations provided by expert food authorities, participants preferred data-driven evidence produced through their dietary self-experiments and peer-to-peer troubleshooting. In this way, the diet corresponds with the Quantified Self practice of self-discovery and the broader self-biohackers' approach to making sense of their health through DIY trial-and-error methods.

However, this libertarian rhetoric of self-care fell short when it came to participants' ability to take full responsibility for the potential failures of their 'innovative' yet also risky diets. Participants were often aware of the possible hazards of the CF practice, such as the limited health safety of self-experimentation or the possibility of losing sensitive personal data. Yet, they did not put a substantial effort into resolving or mitigating these problems and were ready to renew their dependencies on expert food/health authorities in case things go wrong. CF therefore illustrates not only the limitations of quantified diet hacking projects but, in the broader sense, also the limitations of libertarianism and techno-solutionism embraced by the Silicon Valley-like innovation centres. The CF study shows that foodpreneurial aspirations to innovate food systems by turning food practices into quantified data-driven events are limited when implemented in everyday-life contexts.

That is not to say, however, that the Silicon Valley food-tech efforts are wrong and should be discontinued. I have shown that digital food technologies and data-driven practices can support meaningful knowledge exchanges and make people feel better about their diets, bodies and health. The challenge for food-tech innovators is to design products that are carefully reflecting the social sensitivities of present food cultures and practices. Rather than churning out techno-solutions that reduce everyday food experiences into quantified datasets, the goal is to support digital food cultures that are participatory and fun as well as safe and transparent.

Acknowledgements

This work was supported by the European Regional Development Fund-Project 'Creativity and Adaptability as Conditions of the Success of Europe in an Interrelated World' (No. CZ.02.1.01/0.0/0.0/16_019/0000734).

Further reading

Biltekoff, C., Mudry, J., Kimura, A.H., Landecker, H., & Guthman, J. (2014). Interrogating moral and quantification discourses in nutritional knowledge. *Gastronomica: The Journal of Food and Culture, 14*(3), 17–26.

Ferreira, F.R., Prado, S.D., de Carvalho, M.C.V.S., & Kraemer, F.B. (2015). Biopower and biopolitics in the field of Food and Nutrition. *Revista de Nutrição, 28*(1), 109–119.

Kera, D., Denfeld, Z., & Kramer, C. (2015). Food hackers. *Gastronomica: The Journal of Critical Food Studies, 15*(2), 49–56.

Lewis, T., & Phillipov, M. (2018). *Food/media: Eating, Cooking, and Provisioning in a Digital World*. Abingdon: Taylor & Francis.

Sharon, T. (2017). Self-tracking for health and the quantified self: Re-articulating autonomy, solidarity, and authenticity in an age of personalized healthcare. *Philosophy & Technology, 30*(1), 93–121.

References

Anderson, T. (2018). *The Art of Health Hacking: A Personal Guide to Elevate Your State of Health and Performance, Stress Less, and Build Healthy Habits that Matter*. New York: Morgan James.

Barthes, R. (1997). Toward a psychosociology of contemporary food consumption. *Food and Culture: A Reader, 2*, 28–35.

Braun, V., & Clarke, V. (2006). Using thematic analysis in psychology. *Qualitative Research in Psychology, 3*(2), 77–101.

CB Insights (2016, January 15). Food tech startups raise a record $5.7B in 2015. *CB Insights Research Portal*. Retrieved from www.cbinsights.com/research/food-tech-funding-2015/

Comber, R., Choi, J.H.-J., Hoonhout, J., & O'Hara, K. (2014). Designing for human–food interaction: An introduction to the special issue on 'food and interaction design'. *International Journal of Human-Computer Studies, 72*(2), 181–184.

CFS – Complete Foods Survey (2018). Retrieved from http://completefoodsurvey.com/

Crunchbase. (2018). *Soylent – Overview.* Retrieved from www.crunchbase.com/organization/soylent-corporation#section-overview

Dolejšová, M., & Kera, D. (2017). Soylent diet self-experimentation: Design challenges in extreme citizen science projects. *Proceedings of the 2017 ACM Conference on Computer Supported Cooperative Work and Social Computing.* ACM, pp. 2112–2123.

Dolejšová, M., Khot, R.A., Davis, H., Ferdous, H.S., & Quitmeyer, A. (2018). Designing recipes for digital food futures. *Extended Abstracts of the 2018 CHI Conference on Human Factors in Computing Systems*, ACM, pp. 1–8.

FAO – Food and Agricultural Organization of United Nations. (2017). *Food wastage: Key facts and figures.* Retrieved from www.fao.org/news/story/en/item/196402/icode/

Fotopoulou, A., & O'Riordan, K. (2017). Training to self-care: Fitness tracking, bioped-agogy and the healthy consumer. *Health Sociology Review, 26*, 54–68.

Fumey, G., Jackson, P.A., & Raffard, P. (2016). Interview with Richard C. Delerins. High-Tech and Food 2.0. *Anthropology of Food, 11.* Retrieved from https://journals.openedition.org/aof/8109

Ganglbauer, E., Fitzpatrick, G., Subasi, Ö., & Güldenpfennig, F. (2014). Think globally, act locally: A case study of a free food sharing community and social networking. *Proceedings of the 17th ACM Conference on Computer Supported Cooperative Work & Social Computing*ACM, pp. 911–921.

Gioia, T. (2015). San Francisco during the great food awakening. *Virginia Quarterly Review, 91*(3), 250–255.

Jarvis, A. (2014, June 10). Biohackers are turning meals into math across the world. *Medium.* Retrieved from http://munchies.vice.com/en_us/article/kbxg4x/biohackers-are-turning-meals-into-math-across-the-world

Kuznetsov, S., Kittur, A., & Paulos, E. (2015). Biological citizen publics: personal genetics as a site of public engagement with science. *Proceedings of the 2015 ACM SIGCHI Conference on Creativity and Cognition*, ACM, pp. 303–312.

Leek, S., Szmigin, I., & Baker, E. (2015). Consumer confusion and front of pack (FoP) nutritional labels. *Journal of Customer Behaviour, 14*(1), 49–61.

Lupton, D. (2014). Self-tracking modes: Reflexive self-monitoring and data practices. Retrieved from https://papers.ssrn.com/sol3/papers.cfm?abstract_id=2483549

Lupton, D. (2018a). Cooking, eating, uploading: Digital food cultures. In K. LeBesco & P. Naccarato (Eds.), *The Handbook of Food and Popular Culture* (pp. 66–79). London: Bloomsbury.

Lupton, D. (2018b). 'I just want it to be done, done, done!' Food tracking apps, affects, and agential capacities. *Multimodal Technologies and Interaction, 2*(2). Retrieved from www.mdpi.com/2414-4088/2/2/29/htm

Luzar, C. (2013, October 22). Crowdfunding darling Soylent nets $1.5 million in VC funding. *Crowdfund Insider.* Retrieved from www.crowdfundinsider.com/2013/10/24870-crowdfunding-darling-soylent-nets-1-5-million-vc-funding/

Miles, C., & Smith, N. (2015). What grows in Silicon Valley. In H. L. Davis, K. Pilgrim & M. Sinha (Eds), *The Ecopolitics of Consumption: The Food Trade* (pp. 119–137). London: Lexingham Books.

Musk, K. (2016). Why food is the new internet. *Medium.* Retrieved from http://medium.com/food-is-the-new-internet/why-food-is-the-new-internet-4e87810e24a2

Nguyen, H., Chen, S., & Mukherjee, S. (2011). Counter-stigma and achievement of happiness through the freegan ideology. *Association for Consumer Research*. Retrieved from http://www.acrwebsite.org/volumes/v38/acr_v38_16251.pdf

Rhinehart, R. (2013, February 13). How I stopped eating food. *Mostly Harmless*. Retrieved from http://robrhinehart.com/?p=298

Scrinis, G. (2013). *Nutritionism: The Science and Politics of Dietary Advice*. New York: Columbia University Press.

Sneijder, P., & Te Molder, H. (2009). Normalizing ideological food choice and eating practices. Identity work in online discussions on veganism. *Appetite*, *52*(3), 621–630.

Sovijärvi, O., Arina T., & Halmetoja, J. (2016). *Biohacker's Handbook*. Helsinki: Biohacker Center.

Sparrow, J. (2014, May 19). Soylent, neoliberalism and the politics of life hacking. *Counterpunch*. Retrieved from www.counterpunch.org/2014/05/19/solyent-neoliberalism-and-the-politics-of-life-hacking/

Spiteri Cornish, L., & Moraes, C. (2015). The impact of consumer confusion on nutrition literacy and subsequent dietary behavior. *Psychology & Marketing*, *32*(5), 558–574.

Stewart-Knox, B., Rankin, A., Kuznesof, S., Poínhos, R., de Almeida, M.D.V., Fischer, A., & Frewer, L.J. (2015). Promoting healthy dietary behaviour through personalised nutrition: Technology push or technology pull? *Proceedings of the Nutrition Society*, *74*(2), 171–176.

Wiesing, U. (2008). The history of medical enhancement: From restitutio ad integrum to transformatio ad optimum? In B. Gordijn & R. Chadwick (Eds.), *Medical Enhancement and Posthumanity* (pp. 9–24). Dordrecht: Springer.

World Health Organisation (WHO). (2017). *Obesity and overweight factsheet from the WHO*. Retrieved from www.who.int/news-room/fact-sheets/detail/obesity-and-overweight

INDEX

3D food printing 3

abundant eating 90–91
aesthetic labour 60
aesthetic self-tracking 21
affective affordances 37
affirmation ritual 63
affordances 35; affective 37; digital
    technologies 2, 12; networking 157
aging 74, 76
Ahmed, S. 135
alternative food consumption 168
alternative food movement 116
alternative food networks 116, 122–123,
    124–125, 151–152, 168
alternative food politics 162
alternative hedonism 168
amateur restaurant reviewers 8–9, 99–111;
    accuracy 106; authenticity 103, 107–108;
    authority 108–111; comparative analyses
    102; critical engagement 109–110;
    as cultural intermediaries 101, 111;
    descriptors 102; disinterestedness 101,
    106; ethical framework 101; ethics
    105–106; frames 103–104; gastronomic
    field 100–104; methodology 104–105;
    penetration 102; proliferation of 100;
    public identities 99–100
Amazon, food search 1
animal-derived pollutants 75–76
animal rights 92
animal welfare 7, 84
anxiety 30

apps 1–2, 3
Arendt, H. 150
artificial intelligence 155, 185
Arvidsson, A. 116
ascetism 68
Askegaard, S. 30
attention economy, the 120
audiences, social media 35
Australia 118, 151–152, 154; Pete Evans
    case study 121–124
Australian Guideline to Healthy Eating 118
authenticity 9, 85, 92, 103, 107–108

Barad, K. 37
Barnard, A. 131
Barnes, C. 116–117, 119
Bartkowski, J.P. 69
Baumann, S. 103
Beardsworth, A. 69
beautifying applications 21
beauty 21
behaviours, polarisation of 23
Bennett, J. 37
Berthold, D. 131
Big Food 84, 92
binary oppositions 6
binge eating 39–42
biomarkers 188
Black Lives Matter movement 167
blogs 1, 3, 53, 86, 119, 130–131; amateur
    restaurant reviewers 99–111; authenticity
    9, 107–108; comparative analyses 102;
    ethical framework 105–106; food

femininity 132; foodie waste femininity
136–139; food waste posts 135–136;
hyperlinking 130, 135; methodology
104–105; proliferation of 100; public
identities 99–100; Sarah Wilson case
study 133–140
bodies and affects 6
bodily appearance 7
bodily signals, quantification of 183–185
body optimisation 88–89
body, the: carnivalesque 43; ideal 21; images
60, 61; optimisable 187–188; religious
trope 89; social media aesthetics 26–27;
surveillance 20; visualisation 29
Bourdieu, P. 100–101, 103, 111
boyd, d. 58, 156
Braidotti, R. 37
Brixton Village Market 11, 164, 165–167,
168, 170
Broad, G.M. 156, 157
bro culture 42–43
Buttermore, Stephanie 35, 37, 39–42, 44–46

Cairns, K. 35, 119, 132
Caldwell, M. 148, 150–151, 155, 156
calorie counters 44
carnivalesque body, the 43
carnivalesque consumption 6, 35–46
Castells, M. 163–164
Cederström, C. 20
celebrity chefs 5, 114–125; and
contemporary foodscapes 115–117,
119; credibility 115; definition 116–117;
influence 114, 115, 117, 118, 122; media
coverage 118; Pete Evans case study
121–124; and social media 117–125
cheat days 6, 19, 24–26, 31, 35, 36, 39–42
cheat day videos 39–42
Chik, A. 102
choice 150
choice architecture, self-tracking 23
Christopher, A. 69
citizenship 117
civic agriculture 153
Civic Systems Lab 156
Claiborne, C. 101, 104, 106
clean eating 2, 7, 53–65; affirmation ritual
63; definition 54; desserts 59–60; and
gender 60; Glamour Shot 60; and health
60–61; idealisation 62–63, 64; images
59–61; methodological approach 57–59;
and moral emotions 54–56; positive
effects of 60; visual attributes 58–59;
visual content analysis 54
cleanliness 136–137, 139, 140

Cockburn, A. 59–60
cognition 194–195
collaboration, modes of 31
colonialism 136–137
community: Brixton Village Market 164,
165–167, 168, 170; ecosystem 163; and
food mediation 163–173; networked
164–165, 172; network politics 167–169;
and space 172; West Norwood Feast 164,
166, 169–172, **171**, 172
community gardens 153–154
competitive eating 3, 41
Complete Foods diet 12, 193–205;
critiques 199–200; data-sharing
203–204; experiences 198; health issues
202–203; marketplace 196; meals 194;
methodology 197–198; motivations
198–200, 204; online forums 196–197,
201; origins and development 195–197;
peer knowledge sharing 201; self-
experimentation 200–201, 204
connected consumption 149
consumer citizenship 117
consumption choices 30
consumption habits 26
consumption practices 19, 30–31; veganism
8, 82–92
contemporary foodscapes, and celebrity
chefs 115–117, 119
Contois, E. 55, 132
cooking videos 2
corporate wellness schemes 20–21
cosmopolitanisation 165
cravings 89
Crawford, R. 62
creative labour 162
credibility 115
critical content analysis 180
critical discourse 101
culinary capital 169
cultural fields 101–102, 107
cultural intermediaries 101, 111
cultural sociological perspective 71
culture, definition 71

Dark Mountain project 169
datafication 181–182, 188, 193
data-free celebrity science 114
data philanthropy 24, 27–28
data security 203–204
data sharing 31, 203–204
Davies, A. 153, 154, 157
de Certeau M. 189
decluttering 137
De Solier, I. 3, 99, 118, 163, 168–169

desserts, clean eating 59–60
dietary guidelines 199
dietary styles 7
diet literature, language 55
diet personalisation 194
diets 83; restricted 2, 7
digital discourse analysis 71
digital disparities 152
digital food activism 116
digital food cultures: definition 2; diversity
    2; previous research 3–5
digital food technologies 179–189; as
    aspirational forces 187–188; business
    logic 181–182; closed technical systems
    180; Complete Foods diet 193; critical
    sociocultural perspective 181; designers
    179; developers 179; emergence of 181;
    further research 189; Habit 183–185,
    188; HAPIfork 186–187; and healthy
    eating 179; marketing strategy 189;
    metric culture 181; nutrigenomic
    service 183–185, 187; personalized food
    183–185; research 180–181; SmartPlate
    185–186, 188; smart tableware 185–187;
    technological solutionism 181; types 182;
    value proposition 188
digital identities 11
digital inclusion 157
digital media 1
digital technologies 12, 179–189;
    affordances 2, 12
direct-to-consumer (DTC)
    genomics 183
dirt avoidance 55–56
discipline 137, 140
discrimination 157
disgust 43, 56
doggy bag use 131
do-it-yourself ethos 170
Douglas, M. 55–56, 64
dumpster diving 131
Durkheim, E. 56, 64
Dyer, R. 133–134

Eastern mysticism 89
Eat, Cook, Grow (Choi, Foth, & Hearn) 5
eating: accurate way 188–9; as gendered
    activity 90
Edwards, F. 153, 154
Elias, A.S. 21
Eli, K. 116
embodied intimacy 165
empowerment 23, 29
entrepreneurial femininity 40
environmental sustainability 108

Epic Meal Time channel 6, 35, 37,
    42–43, 44–46
equity 150
ethical consumption 7
ethics 31; amateur restaurant reviewers
    105–106; professional restaurant
    reviewers 105
Evans, Pete 9, 114–115, 121–124; *Bubba
    Yum Yum* 115, 121, 122; Facebook page
    121, 122; foodscape brand 122; influence
    122; positioning 123–124; qualifications
    122; social media followers 122
exclusion 149, 150, 152, 157
experts and expertise 8–10, 118, 120; and
    institutionalised knowledge 124; Pete
    Evans case study 121–124

Facebook 2, 19, 35, 118, 119, 121, 135,
    147, 171
Fearnley-Whittingstall, Hugh 135–136
Feenstra, G. 151
femininity 9, 40, 129–140; control of 140;
    digitally mediated 131–132; food 129–
    140; foodie waste 136–139, 140; and
    food waste 132–140; idealised 132, 134;
    methodology 133–134; performance
    132; racialised 130; Sarah Wilson case
    study 133–140; white 132; whiteness
    133–135
Ferreri, M. 168
field theory 111
fitspo cultures 40
food: abandonment of 199–200, 204;
    appearance 26–27; definitions 83;
    and health 76–77; media portrayals 1;
    medicalised 78; and moral emotions
    54–56; and morality 54–56, 83;
    multisensory pleasures 6; power 35; roles
    82; symbolic representation 53–65; on
    YouTube 38–39
food activism 4, 116, 149, 162
food apps 147–148
food-based identities 114
food celebrities 116
food choices 30, 195; individual
    responsibility 76
food consumption: abundant eating 90;
    Buttermore 35–36, 37, 39–42, 44–46;
    carnivalesque 35–46; cheat day videos
    39–42; discussion 43–46; Epic Meal
    Time channel 35–36, 37, 42–43, 44–46;
    excessive 6, 35–46; gender norms 45;
    self-monitored 6; sociocultural discourses
    44; wrong 6; on YouTube 38–39
food culture 99

food femininity 129–140; blogs 132; control of 140; digitally mediated 131–132; literature 132–133; methodology 133–134; racialised 130; Sarah Wilson case study 133–134; whiteness 133–135
food futures 11–12
food governance 117
food groups 82
food hackathons 155
foodies 99, 111, 169
foodie waste 136–139
foodie-waste femininity 9, 129–140
food information, competing 199
food insecurity 149, 150
food justice 10, 148, 150, 156, 157, 172
food knowledge, production of 119
food literacy 153
food mediation: biopolitical forces 162; Brixton Village Market 164, 165–167, 168, 170; and community 163–173; ecosystem 163; networked 164–165, 172; network politics 167–169; social media 162–173; and space 162, 165–167, 172; West Norwood Feast 164, 166, 169–172, **171**
food miles 107–108
Foodmunity 148
FoodPanda 147
food politics 85, 172
food porn 4, 6, 45, 59–60, 132
food preferences 12
foodpreneurs 193
food production 10
food-related domains 35
food rescue tactics 130, 131, 138–139
foodscapes 116, 150; alternative food networks 151–152; connections and tensions 150–152; corporate capture 150; local 151–152, 157; neighbourhood level 155–156; participatory 147–157; participatory ecosystems 153–156; voice 149–150, 157
food-sharing 153, 154
food videos: agential capacities 44; Buttermore 35–36, 37, 39–42, 44–46; carnivalesque 35–46; cheat day videos 39–42; discussion 43–46; Epic Meal Time channel 35–36, 37, 42–43, 44–46; gender norms 45; Google's analysis 39; popularity 39; vicarious pleasure 41; YouTube 38–39
food waste: digitally mediated 133; and femininity 132–140; household studies 132–133; and race 131

food waste reduction 9, 129–140; food rescue tactics 138–139; methodology 133–134; minimalism 137; Sarah Wilson case study 133–140; whiteness 133–135
Foucault, M. 20
Fraser, C. 132–133

gastronomic field, the 100–104
gender 3; and clean eating 60; and Instagram 60; norms 12, 45
genetic factors 77
Georgiou, M. 165
Gibbs, N. 183
Gill, R. 21
Giordano, A. 116
Goffman, E. 56, 62, 63, 64
gonzo porn 45
good health 21
Goodman, D. 116
Goodman, M.K. 35, 119, 148
Goodyear, V.A. 29
Grasseni, C. 153
grassrooting 163–164
gratification 23
Greenebaum, J. 70
greenlining 166
Gregory, A. 24, 26
guilt 20, 23, 26, 29, 30

Habit 183–185, 188
habitus 100–101, 111
HAPIfork 186–187
Haraway, D. 37, 38
Hari, Vani 119
Harvey, D. 165
Haverda, T. 69
Hay, Donna 114
health 88; and clean eating 60–62; desire for 31; discourses of individual responsibility 76–77, 202–203; environmental factors 77; and food 76–77; genetic factors 77; good 21; idealised 77; individualised conceptions of 61–62; and Instagram 60–62; as key marker of value 76; risks to 77; self-management 28; structural factors 77
health choices 23
health expertise 31
health identity 19, 23, 26–27, 29; morals and values 20
health injunctions, disregard for 46
healthism 7–8, 82; and animal-derived pollutants 75–76; cultural sociological perspective 71; definition 69; digital

discourse analysis 71; discourses of individual responsibility 76–77; health veganism 69; holistic veganism 69; lifestyle factors 69; methodology 71–72; religious themes 69–70; and veganism 7, 68–78; veganism-related discussions 72–75
health management 21
health transformation narratives, vegans 73–75
health veganism 69
healthy eating 82–84; and digital food technologies 179
healthy living, representations of 24
Heyes, C.J. 31
Hine, Dougald 166, 167, 169
holistic veganism 69
household food waste studies 132–133
household food waste work 133
human–computer interaction 5
human–food automation 193–194
human–food interaction 5
Human Genome Project 183
hyper-femininities 6
hyperlinking 130, 135, 138–139
hyper-masculinities 6

idealisation 62–63, 64
identity 82
identity-formation 99
illness, morals and values 20
immorality 29–30
inclusion 147–157; barriers to 153; connections and tensions 150–152; digital 157; locatedness 150; neighbourhood level 155–156; participatory ecosystems 153–156; voice 149–150, 157
individual responsibility 76–77
information overload 120
Instagram 2, 3, 19, 35, 85, 147, 163, 169; affirmation ritual 63; Before/After shot 58, 60–1; body images 60, 61; clean eating 7, 53–65; and gender 60; Glamour Shot 58, 60; hashtags 57–58, 61–62; and health 60–61; and idealisation 62–63, 64; images 59–61; interface 57; Kissy Face 58, 60; methodological approach 57–59; Muscle Presentation 59, 61; Nature Shot 59; profiles 57; ritualised interactions 63; trending 58; visual attributes 58–59; visual content analysis 54
institutionalised knowledge 124
Intelligent Nutrition Platform 185

Jarvis, A.-A. 199
Johnston, J. 35, 103, 116, 119, 132
justice 77

Keil, T. 69
Kristensen, D.B. 19, 30
Kumar, R. 153

lad culture 42–43
Lavis, A. 39
Lefevre, H. 165
leftovers 130, 131, 138–139
Levy, P. 99
Lewis, T. 147, 149, 154, 162, 169, 171
lifestyle choices 32
lifestyle corrections 20
lifestyle factors, healthism 69
lifestyle influencers 4, 9–10
lifestyle media 168
Lim, M. 30
living well 89
local food movement 151
local foodscapes 151–152, 157
locatedness 150
Luise, V. 116
Lupton, D. 20, 21, 204
Lyson, H.C. 116

McFarlane, C. 156
Mail Online 130, 138–139
major retailers, domination of 151
Manzini, E. 157
marketing 11–12
market orientations 163
masculinity 3, 46
Mattila, M. 133
meals, curating 27
meat 70
mechanical approach 29–30
mediated foodscapes 35
mental health 31, 88
meokbang 3
Mesiranta, N. 133
metric culture 181
micro-celebrities 4, 9, 85, 92
Miles, C. 199
Mills, E. 85
Moore, P. 20
moral eating 92
moral emotions, and food 54–56
moral 'healthy' subject, performing 22–24
morality 20, 29, 137; and food 54–56, 83
moral obligation 30

Närvänen, E. 133
Nash, M. 134
neoliberalism 20, 24, 29, 69, 78, 84, 92, 117
network activism 166
networked publics 156
networked society, the 163–165, 172–173
networking affordances 157
network politics 167–169
networks 11
Nielsen 115, 121–124
normative ideals 7
Norwood & Brixton Foodbank 172
nutrigenomic service 183–185, 187
nutritionism 199–200
nutrition labels 199

obesity crisis 83
Oliver, Jamie 5, 9, 42, 114, 118–119
online consumer reviews 100
orthorexia nervosa 83
Ortiz, A. 185

paleo diet, the 114, 115, 121–122
Paltrow, Gwyneth 87
Pantzar, M. 181
Parizeau, K. 132–133
participatory culture 1, 85
participatory ecosystems 153–156
participatory foodscapes 147–157;
    alternative food networks 151–152;
    barriers to 153; connections and
    tensions 150–152; local 157; locatedness
    150; neighbourhood level 155–156;
    participatory ecosystems 153–156; voice
    149–150, 157
participatory sharing 38
peer knowledge sharing 201
personalized food 183–185, 194
personal responsibility 61
Phillipov, M. 85, 147, 167–168
place, attachments to 150
pleasure 43, 89–91
political participation 153
Pollan, M. 83
pop-ups 168, 170
postfemininity 3
Prigge, C. 19
privacy 203–204
privatisation 165
professional restaurant reviewers 104;
    authenticity 107, 108; ethics 105
pro-localism 170
psychedelic microdosing 194–195
public health 82, 85
public space 53

purity 55–56, 76, 131, 139, 140
Purpura, S. 28

quantification 195

race 129; and cleanliness 136–137; and food
    waste 131
racial relations 153
Ramachandran, D. 118
Ramsay, Gordon 42
regulated eating, tableware for 185–187
regulatory design tools, and self-tracking 23
religiosity 55
religious themes 69–70
restaurant reviewing: accuracy 106; amateur
    8–9, 99–111; authenticity 103, 107–108;
    authority 108–111; critical engagement
    109–110; as cultural intermediaries 101,
    111; descriptors 102; disinterestedness
    101, 106; ethical appraisal 8; ethical
    framework 101; ethics 105–106;
    frames 103–104; gastronomic field
    100–104; hierarchy 111; magazines 103;
    methodology 104–105; professional 9,
    104, 105, 107, 108
restricted diets 2, 7
Rhinehart, R. 195–196, 198
ritual 54–55, 56, 62, 63, 64
Robinson, A. 20
robotics 155
Rodney, A. 132
Rolls-Royce 182
rootedness 164
Rosa Labs 196
Roth-Gordon, J. 137
Rousseau, S. 102, 114, 118
Ruckenstein, M. 181

Saldanha, A. 129
self-betterment discourses 24, 29
self-biohacking 195
self-care 202–203, 204
self-discipline 20, 26, 31, 61
self-discovery 184
self-experimentation 200–201, 204
self-governance 31
self-improvement 92
self-knowledge 22–23
self-management 28, 29
self-perfectibility 30
self-presentation 56
self-regulation 29
self-responsibility 7, 12
self-tracking 6, 19–32, 180, 181; aesthetic
    21; background 20–21; burdens of

27–29, 32; cheat days 19, 24–26, 31; choice architecture 23; consumption practices 20; data philanthropy 24; data security 203–204; data-sharing 19; dehumanises 28; and empowerment 23, 29; goals 28; gratification 23; and judgement 23–24; mechanical approach 29–30; methodology 21–22; multiple devices 28; obsessive 30; performing the moral 'healthy' subject 22–24; and regulatory design tools 23; regulatory frameworks 31, 32; and the self 19; self-betterment discourses 24; self-discipline 26; and social media aesthetics 26–27; social media norms 28–29; technologies 19
self-transformation 20
self-worth 32
sensory pleasure 8
servitisation 182
sexism 6, 43
shame 21, 29, 30
sharing ethos 85
Silicon Valley, California 11–12, 193–205
simplicity 137
Slocum, R. 129, 170
small-scale food businesses, economic viability 152
SmartPlate 185–186, 188
smart tableware 185–187, 188
Smith, N. 199
Sneijder, P. 75
social bonding 200
social consciousness 9
social disengagement 149
social fitness 21
social influencers 120
social justice 148, 170
social media 2, 3, 21, 84–85; audiences 35; authenticity 9; and celebrity chefs 17–25; food mediation 162–173; participatory engagements 36; platform capital 169; prevalence of 53; rapid expansion 124; self-tracking platforms 19
social media aesthetics 26–27
social media influencers 4, 5, 9; Pete Evans case study 121–124
social media norms, self-tracking 28–29
Soylent diet 195–196
space, and food mediation 162
Space Makers 164, 166–167, 169
space making 164
spatialities and politics 10–11
Spicer, A. 20
status relations 56

Stebbins, R.A. 99
Stewart-Knox, B. 203
Stewart, Martha 114
superfoods 63
supply chains 151
sustainability 4, 10, 151, 169
Sutinen, U.-M. 133

Talbot, V.C. 21
technological affordances 2
technological solutionism 181, 194, 205
technology developers 179
Te Molder, H.F.M. 75
thing-power 37
Thompson, C.J. 43
truth discourses 119
Turner, B. 155
Twitter 4–5, 35, 119, 135

United States of America, alternative food movement 116
urban agriculture 153
urban gardens 169
urban imaginaries 164–165
urbanization 165
urban strategy 173

value creation 181–182
values 20
Vásquez, C. 102
vegans and veganism 2, 3; as abundance and pleasure 89–91; abundant eating 90; activist position 92; and aging 74, 76; and animal-derived pollutants 75–76; consumption practices 8, 82–92; constructions of 87; cultural sociological perspective 71; darker side 70; diets 73–75; digital discourse analysis 71; discourses of individual responsibility 76–77; engagement with 68; for health and wellbeing 87–89; health benefits 7, 68–78; health problems 73; health-related discussions 72–75; health transformation narratives 74–75; health veganism 69; healthy eating 82–84; holistic veganism 69; hostility towards 68, 70; increase in 84; as learned skill 91–92; literature 69–70; and masculinity 3; methodology 71–72; moral eating 92; motivation 69–70; negative changes 74–75; numbers 68; online forums 68; perceptions of 68, 69, 72–73; personal experience 74; religious themes 69–70, 89; themes of renewal and rebirth 74–75, 77; victim-blaming 77; visible 84

vegan studies 84
vegan vlogs 8, 82–92; abundance and
    pleasure themes 89–91; abundant eating
    90; body optimisation 88–89; content
    86; learned skill themes 91–92; lessons
    87–92; methodological approach 87;
    moral eating discourse 92; numbers
    86; performative position 91; physical
    health/mental health claims 88; style 86
VegChat 71–75, 76, 77–78
Veit, H. 139
vicarious pleasure 41
victim-blaming 77
visioning 172
visualisation 29
vlogs 38; definition 86; lessons 87–92;
    numbers 86; performative position 91;
    vegan 8, 82–92
voice 149–150, 157

websites 1
weight loss 88, 89
Weight Watchers 188
wellbeing 82

wellness 20, 83
wellness diets 83, 89
West Norwood Feast 11, 164, 166,
    169–172, **171**
whiteness 133–134, 134–135, 166
Wilson, Sarah 9–10, 129; blog 130; case
    study 133–140; femininity 130–131;
    foodie waste femininity 136–139; food
    rescue tactics 130, 131, 138–139; food
    waste posts 135–136; homepage 134–
    135; methodology 133–134; whiteness
    133–135
Witthall, E. 139

Yeatman, Heather 122
YouTube 2, 5, 6, 85, 169; Buttermore
    35–36, 37, 39–42, 44–46; cheat day
    videos 39–42; discussion 43–46; Epic
    Meal Time channel 35–36, 37, 42–43,
    44–46; food on 38–39; food videos
    35–46; gender norms 45; participatory
    engagements 36; participatory sharing
    38; subscribers 38, 40, 42; vegan
    vlogs 8, 82

Printed in the United States
by Baker & Taylor Publisher Services